Modelling with AutoCAD Release 14

for Windows NT and Windows 95

Other titles from Bob McFarlane

Beginning AutoCAD   ISBN 0 340 58571 4

Progressing with AutoCAD   ISBN 0 340 60173 6

Introducing 3D AutoCAD   ISBN 0 340 61456 0

Solid Modelling with AutoCAD   ISBN 0 340 63204 6

Starting with AutoCAD LT   ISBN 0 340 62543 0

Advancing with AutoCAD LT   ISBN 0 340 64579 2

3D Draughting using AutoCAD   ISBN 0 340 67782 1

Beginning AutoCAD R13 for Windows   ISBN 0 340 64572 5

Advancing with AutoCAD R13 for Windows   ISBN 0 340 69187 5

Modelling with AutoCAD R13 for Windows   ISBN 0 340 69251 0

Using AutoLISP with AutoCAD   ISBN 0 340 72016 6

Beginning AutoCAD R14 for Windows NT and Windows 95   ISBN 0 340 72017 4

Advancing with AutoCAD R14 for Windows NT and Windows 95   ISBN 0 340 74053 1

Modelling with AutoCAD Release 14

for Windows NT and Windows 95

Bob McFarlane
MSc, BSc, ARCST, MIMechE, MIEE, MIED, MILog

CAD Course Leader, Motherwell College,
AutoDESK Educational Developer

ARNOLD

A member of the Hodder Headline Group
LONDON • SYDNEY • AUCKLAND
Copublished in North, Central and South America
by John Wiley & Sons Inc.
New York • Toronto

To Helen with love

First published in Great Britain 1999 by
Arnold, a member of the Hodder Headline Group,
338 Euston Road, London NW1 3BH

http://www.arnoldpublishers.com

Copublished in North, Central and South America by
John Wiley & Sons Inc., 605 Third Avenue
New York, NY 10158-0012

Whilst the advice and information in this book is believed to be true
and accurate at the date of going to press, neither the author nor the publisher
can accept any responsibility or liability for any errors or omissions
that may be made.

British Library Cataloguing in Publication Data
A catalogue record for this book is available from the British Library

Library of Congress Cataloging-in-Publication Data
A catalog record for this book is available from the Library of Congress

ISBN 0 340 73161 3
ISBN 0 470 32898 3 (Wiley)

2 3 4 5 6 7 8 9 10

Commissioning Editor: Sian Jones
Project Manager: Robert Gray
Production Editor: Liz Gooster
Production Controller: Sarah Kett
Cover design: Terry Griffiths

Produced by Gray Publishing, Tunbridge Wells, Kent
Printed and bound in Great Britain by The Bath Press, Bath
and The Edinburgh Press Ltd, Edinburgh

What do you think about this book? Or any other Arnold title?
Please send your comments to feedback.arnold@hodder.co.uk

Contents

Preface

This book is intended for the Release 14 user who wants to learn about modelling. My aim is to demonstrate how the user can create 3D wire-frame models, surface models and solid models with practical exercises backed up by user activities. The concept of creating multiple viewports will also be discussed, as will two new Release 14 modelling concepts: the VIEW and DRAW commands.

This book will provide an invaluable aid to a wide variety of users, ranging from the capable to the competent. The book will assist students on any national course which requires 3D draughting and solid modelling, such as City & Guilds, BTEC and SQA as well as students at higher institutions. Users in industry will find the book useful as a reference and an 'inspiration'.

Reader requirements

The following are the requirements I consider important for using the book:

- the ability to draw with Release 14
- the ability to use icons and toolbars
- an understanding of how to use dialogue boxes
- the ability to open and save drawings to a named folder
- a knowledge of model/paper space would be an advantage, although this is not essential.

Using the book

The book is essentially a self-teaching package with the reader working interactively through exercises using the information supplied. The various prompts and responses will be listed in order and dialogue boxes will be included where appropriate.

The following points are important:

- All drawing work should be saved to a named folder. The folder name is at your discretion but I will refer to it as R14MOD, e.g. open drawing R14MOD\MODEL1.
- Icons will be displayed when the command is used for the first time.
- Menu bar selection will be in bold type, e.g. **Draw–Surfaces–3D Face**.
- Keyboard entry will also be in bold type, e.g. **VPOINT**, **UCS**, etc.
- Prompts will be in typewriter type, e.g. `First corner`.
- The symbol <R> will require the user to press the return/enter key.

Note

All the exercises and activities have been completed using Release 14. I have tried to correct any errors in the drawings and text, but should any mistakes occur, I apologize for them and hope they do not spoil your learning experience. Modelling is an intriguing topic and should give you satisfaction and enjoyment.

Any comments you have about how to improve the material in the book would be greatly appreciated.

Chapter 1

The 3D standard sheet

To assist us with the models which will be created, a standard sheet (prototype drawing) will be made with layers, a text style, dimension styles, etc. This standard sheet will be saved as a template.

1 Start AutoCAD R14 and:
   prompt   Start Up dialogue box
   respond   **pick Use a Wizard**
   prompt   Use a Wizard dialogue box
   respond   **pick Advanced Setup then OK**
   prompt   Advanced Setup dialogue box
   respond   select the following to the various steps:
               Step 1   Units: Decimal; Precision 0.00; Next>>
               Step 2   Angle: Decimal Degrees; Precision 0.0; Next>>
               Step 3   Angle Measurement: East(0); Next>>
               Step 4   Angle Direction: Counter-Clockwise(+); Next>>
               Step 5   Area: Width 420 × Length 297 (i.e. A3); Next>>
               Step 6   Title Block
                         Title Block Description: No Title Block
                         Title Block File Name: None
                         Next>>
               Step 7   Layout
                         Paper Space layout capability: No
                         Done.

2 A blank screen will be displayed.

3 *Layers*
   Menu bar with **Format–Layer** and make the following new layers:

name	*colour*	*linetype*
MODEL	RED	continuous
TEXT	GREEN	continuous
DIM	MAGENTA	continuous
OBJECTS	BLUE	continuous
SECT	number: 96	continuous
0	white	continuous

   Note: other layers will be added as required.

4 *Text style*
   Menu bar with **Format–Text Style** and make a new text style:
   Name: ST1
   Font: romans.shx
   Height: 0; Width factor: 1; Obliquing angle: 0.

5 *Units*
   Menu bar with **Format–Units** and:
   Units: Decimal with Precision: 0.00
   Angle: Decimal with Precision: 0.0.

6 *Limits*
   Menu bar with **Format–Drawing Limits** and:
   prompt   Lower left corner and enter: **0,0 <R>**
   prompt   Upper right corner and enter: **420,297 <R>**.

7 *Drawing Aids*
   Menu bar with **Tools–Drawing Aids** and set:
   Blips: off
   Highlight: on
   Snap: 5 and grid: 10 – not generally used in 3D
   Other aids as required.

8 *Dimension style*
   Menu bar with **Dimension–Style** and:
a) Name:	change to **3DSTD** and pick Save
b) Geometry:	Spacing: 10
	Extension: 2.5
	Origin Offset: 2.5
	Arrowheads: both Closed Filled
	Size: 3.5
	Centre: None
c) Format:	User defined, Force Line Inside: Both OFF
	Fit: Best Fit
	Horizontal Justification: Centered
	Vertical Justification: Above
	Text: Inside Horizontal: OFF – no tick
	Outside horizontal: ON – tick in box
d) Annotation:	Units: Decimal; Precision: 0.00
	Trailing: ON, i.e. tick in box
	Decimal Angle; Precision: 0.0
	Trailing: ON
	Tolerance: None
	Enable Units: OFF, i.e. no tick
	Text: Style: ST1
	Height: 3.5
	Gap: 1.5

   e) Save to 3DSTD.

9 Make layer 0 current and menu bar with **Draw–Rectangle** and:
   prompt   First corner and enter: **0,0 <R>**
   prompt   Other corner and enter: **420,290 <R>**
   This rectangle will save as a 'reference base' for our models.

10 Zoom–all and pan to suit.

11 Make layer MODEL current.

12 Menu bar with **File–Save As** and:
  prompt   Save Drawing As dialogue box
  respond   1. scroll at Save as type
             2. pick **Drawing Template File (*.dwt)**
             3. enter File name as: **3DSTDA3**
             4. pick Save
  prompt   Template Description dialogue box
  respond   1. enter: **This is my 3D standard sheet**
             2. pick OK.

13 The created standard sheet has been saved as a template file with the name 3DSTDA3. It has been added to the existing list of AutoCAD templates in the Template folder.
  Note: *a*) we could have saved the template file in our R14MOD folder. You can re-save the standard sheet in this folder if required
      b) saving the standard sheet as a template will stop the user 'inadvertently' saving any drawing with the 3DSTDA3 name.

14 *Checking the standard sheet*
  Menu bar with **File–Open** and:
  prompt   Select File dialogue box
  respond   1. scroll at Files of type
             2. pick **Drawing Template File(*.dwt)**
  prompt   Template folder with list of template files
  respond   1. **pick 3DSTDA3**
             2. pick Open.

15 The created standard sheet will be displayed and is ready for use.

16 Step 14 is the method by which all new drawings will be started, i.e. with the 3DSTDA3 template file.

Chapter 2

Extruded 3D models

An extruded model is created by extruding a 'shape' upwards or downwards from a horizontal plane – called the ELEVATION plane. The actual extruded height (or depth) is called the THICKNESS and can be positive or negative relative to the set elevation plane. This extruded thickness is **always perpendicular** to the elevation plane. The basic extruded terminology is displayed in Fig. 2.1.

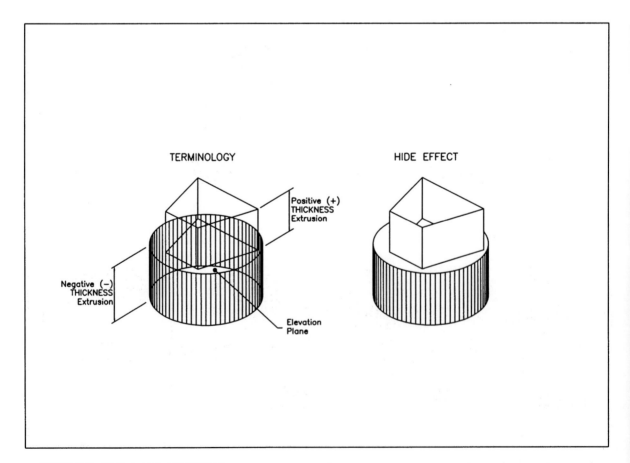

Figure 2.1 Basic extruded terminology.

Extruded example 1

The example is given as a series of steps in Fig. 2.2, so open your 3DSTDA3 template file and display toolbars to suit, e.g. Draw, Modify and Object Snap. Layer MODEL should be current?

Step 1: the first elevation

1 At the command line enter **ELEV <R>** and:
   prompt   New current elevation<0.00> and enter: **0 <R>**
   prompt   New current thickness<0.00> and enter: **50 <R>**.

2 Nothing appears to have happened?

3 Select the LINE icon and draw:
   From point **40,40 <R>**
   To point   **@100,0 <R>**
   To point   **@100<90 <R>**
   To point   **@−100,0 <R>**
   To point   **C <R>** – the close option.

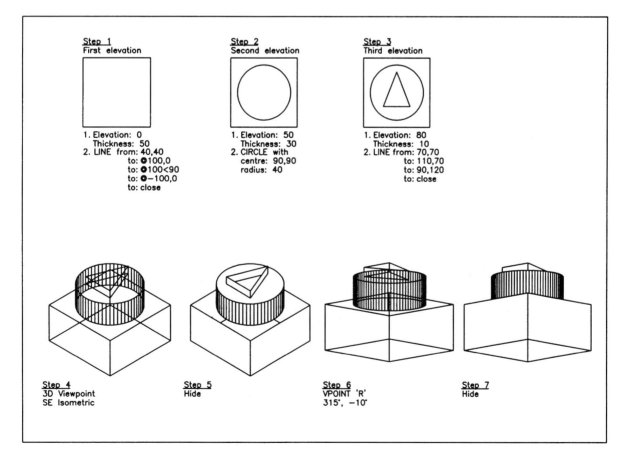

Figure 2.2 Extruded example 1.

Step 2: the second elevation

1 At the command line enter **ELEV <R>** and:
   prompt   New current elevation<0.00> and enter: **50 <R>**
   prompt   New current thickness<50.00> and enter: **30 <R>**.

2 Select the CIRCLE icon and:
   a) centre point: **90,90 <R>**
   b) radius: **40 <R>**.

Step 3: the third elevation

1 With the ELEV command:
   a) set the current elevation to 80
   b) set the current thickness to 10.

2 With the LINE icon, draw:
   From point **70,70 <R>**
   To point   **110,70 <R>**
   To point   **90,120 <R>**
   To point   **C <R>**.

3 Now have a triangle inside a circle inside a square, and appear to have a traditional 2D plan type drawing. Each of the three shapes has been created on a different elevation plane:
   a) square: elevation 0
   b) circle: elevation 50
   c) triangle: elevation 80.

Step 4: viewing the model in 3D

To 'see' the model in 3D the 3D Viewpoint command is required, so:

1 From the menu bar select **View–3D Viewpoint–SE Isometric**.

2 The model will be displayed in 3D. The black 'drawing border' is also displayed in 3D and acts as a 'base' for the model.

3 The orientation of the model is such that it is difficult to know if you are looking down on it, or looking up at it. This is common with 3D modelling and is called **ambiguity** and another command is required to 'remove' this ambiguity.

4 At this stage save your model with **File–Save As** and ensure:
   a) File type is: AutoCAD R14 Drawing (*.dwg)
   b) Save in: R14MOD – your named folder
   c) File name: **EXT_1** – the drawing name.

5 This saves the drawing as **R14MOD\EXT_1.dwg**.

Step 5: the hide command

1 From the menu bar select **View–Hide** and the model will be displayed with hidden line removal and is now easier to 'see'.

2 From the screen display it is obvious that the model is being viewed from above, but it is possible to view from different angles.

3 Menu bar with **View–Regen** to 'restore' the original model.

Step 6: another viewpoint

1 At the command line enter **VPOINT <R>** then **R <R>** and:

prompt   Enter angle in XY plane from X-axis and enter: **315 <R>**
prompt   Enter angle in XY plane and enter: **–10 <R>**.

2 The model will be displayed from a different viewpoint without hidden line removal.

Step 7: the hide command

1 At the command line enter **HIDE <R>**.

2 The model will be displayed with hidden line removal and is being viewed from below.

3 At the command line enter **REGEN <R>** to restore the original.

Task

1 Restore the 3D Viewpoint–SE Isometric.

2 Note that the square and triangular shapes have no top or bottom while the circle shape has. This is always the case with extruded models.

3 With the ERASE command pick any line of the 'base' – a complete 'side' is erased because it is an extrusion.

4 Undo the erase effect with **U <R>**.

5 Using the erase command pick any point on the top 'circle' and the complete 'cylinder' will be erased.

6 Undo this erase effect.

7 This completes our first extrusion exercise.

8 *Note*: although Fig. 2.2 displays several different viewpoints of the model on 'one sheet' this concept will not be discussed until a later chapter. At present you will only display a single viewpoint of the model.

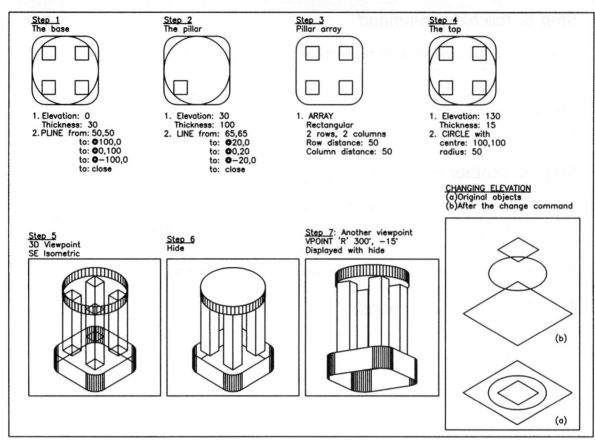

Figure 2.3 Extruded example 2.

Extrusion example 2

Open your 3DSTDA3 template file, layer MODEL current and refer to Fig. 2.3.

Step 1: the base

1 With the ELEV command set to the current elevation to 0 and set the current thickness to 30.

2 With the polyline icon from the Draw toolbar, draw a 0 width polyline:
   From point   **50,50 <R>**
   To point   **@100,0 <R>**
   To point   **@0,100 <R>**
   To point   **@−100,0 <R>**
   To point   **C <R>**.

3 Menu bar with **Modify–Fillet** and:
   prompt   Polyline/Radius... and enter: **R <R>** – the radius option
   prompt   Enter fillet radius and enter: **20 <R>**.

4 Select the FILLET icon from the Modify toolbar and:
   prompt   Polyline/Radius...
   enter   **P <R>** – polyline option
   prompt   Select 2D polyline
   respond **pick any point on the polyline**.

5 The drawn polyline will be filleted at the four corners.

Step 2: the first pillar

1 Set the elevation to 30 and the thickness to 100.

2 With the LINE icon draw a 20 unit square the lower left corner being at the point 65,65.

Step 3: arraying the pillar

1 Select the ARRAY icon from the Modify toolbar and:
   prompt   Select objects
   respond   **window the square** then right-click
   prompt   Rectangular or Polar and enter: **R <R>** – rectangular
   prompt   Number of rows and enter: **2 <R>**
   prompt   Number of columns and enter: **2 <R>**
   prompt   Unit cell or row distance and enter: **50 <R>**
   prompt   Distance between columns and enter: **50 <R>**.

2 The square is arrayed in a 2 × 2 matrix pattern.

Step 4: the top

1 Set the elevation to 130 and the thickness to 15.

2 Draw a circle, centred on 100,100 with radius of 50.

Step 5: the 3D viewpoint

1 Menu bar with **View–3D viewpoint–SE Isometric**.

2 The model is displayed in 3D but appears rather 'cluttered'.

Step 6: hiding the model

1 Menu bar with **View–Hide** model displayed with hidden line removal.

2 Menu bar with **View–Regen** to restore original.

3 At this stage save the model as **R14MOD\EXT_2**.

Step 7

1 At the command line enter **VPOINT <R>** then:
   a) enter: **R <R>** – the rotate option
   b) enter angles of **300** and **–15**.

2 Hide the display.

   Before leaving this exercise, there are a few new concepts which I would like to introduce.

Task 1

1 Still have the extruded model displayed?

2 Menu bar with **View–3D Viewpoint–Plan View–World UCS**.

3 The model is displayed 'as drawn'.

4 Select the PROPERTIES icon from the Object Property toolbar and:
   prompt   `Select objects`
   respond   **window the four pillars then right-click**
   prompt   Change Properties dialogue box
   respond   1. pick **Color – Select Color** dialogue box
             2. pick **Standard Colour Blue** then **OK**
   prompt   Change Properties dialogue box with Blue Color
   respond   **pick OK**.

5 At the command line enter **CHANGE <R>** and:
   prompt   `Select objects`
   respond   **pick the circle then right-click**
   prompt   `Property/<Change point>`
   enter   **P <R>** – the property option
   prompt   `Change what property...`
   enter   **C <R>** – the color option
   prompt   `New color<BYLAYER>`
   enter   **GREEN <R>**
   prompt   `Change what property...`
   respond   **right-click** – no more properties to change.

6 We now have the following:
   a) the filleted square base: red
   b) the four pillars: blue
   c) the circular top: green.

7 Display at 3D Viewpoint–SE Isometric then hide – interesting?

8 Menu bar with **View–Shade–16 Color Filled** – impressive?

9 Repeat the View–Shade sequence but this time select 256 Color Edge Highlight – any difference from previous shade?

10 Menu bar with View–Regen to 'restore' original.

Task 2

1 Restore the original layout with the sequence View–3D Viewpoint–Plan View–World UCS.

2 With ELEV at the command line, set the elevation and thickness to 0.

3 Draw the following objects to right of the model:
   a) a 100 sided square
   b) a 40 radius circle inside the square
   c) a 30 sided square inside the circle.

4 Menu bar with View–3D Viewpoint–SE Isometric and the three shapes will be displayed at elevation 0.

5   At the command line enter **CHANGE <R>** and:
    a) pick the circle then right-click
    b) enter **P <R>** for the property option
    c) enter **E <R>** for the Elev(ation) option
    d) enter **50 <R>** as the new elevation
    e) right-click to end command.

6   Using the CHANGE command, alter the elevation of the four lines of the small square to 80.

Task 3

Investigate altering the THICKNESS of objects:
a) with the command line CHANGE
b) with the PROPERTIES icon.

The extruded model exercise 2 is now complete and you can proceed to the assignment for this chapter, after reading the summary.

Summary

1   An extruded model is created using an elevation and thickness.

2   Extruded models created from straight lines have no top or bottom surfaces – only sides.

3   The elevation and thickness are set using ELEV from the command line.

4   The CHANGE command can be used to set the elevation and alter the thickness of objects.

5   The PROPERTIES icon can be used to alter the thickness of objects.

6   Extruded models are viewed in 3D with the 3D Viewpoint command which will be discussed in detail in a later chapter.

7   3D models are displayed with AMBIGUITY, i.e. are you looking down from the top or up from the bottom?

8   The HIDE command is used to display 3D models with hidden line removal. This removes the AMBIGUITY effect.

9   The REGEN (regenerate) command restores a 3D model with all lines displayed.

10 The SHADE command gives useful displays with coloured objects. It has four options.

Assignment

One extruded activity has been included for you to attempt (all of the activity drawings are shown at the end of this book).

Activity 1: Half-coupling

Using the reference sizes given create a 3D extruded model of the half-coupling starting with your 3DSTDA3 template file. All the elevation and thickness values are given on the drawing but to help:

a) Coupling base: elevation 0; thickness 40; colour red
b) Bolt: elevation 40; thickness 15; colour blue
c) Shaft: elevation 40; thickness 25; colour green.

When the model has been created, view from:

a) 3D Viewpoint SE Isometric
b) SE isometric with hide
c) Command line VPOINT R with angles of 300 and −5
d) VPOINT R entry with hide.

Also:

a) try some other viewpoint entries
b) SHADE?

Chapter 3

The UCS and 3D coordinates

AutoCAD uses two coordinates systems:
a) the world coordinate system (**WCS**) and
b) the user coordinate system (**UCS**).

The WCS

All readers should be familiar with the basic 2D coordinate concept of a point described as P1 (30,40) – Fig. 3.1. Such a point has 30 units in the positive *X*-direction and 40 units in the positive *Y*-direction. These coordinates are relative to an *XY*-axes system with the origin at the point (0,0). This origin is normally positioned at the lower left corner of the screen and is perfectly satisfactory for 2D draughting but not for 3D modelling.

Drawing in 3D requires a third axis (the *Z*-axis) to enable three-dimensional coordinates to be used. The screen monitor is a flat surface and it is difficult to display a three-axis coordinate system on it. AutoCAD overcomes this difficulty by using an **ICON** and this icon can be moved to different positions on the screen and can be orientated on existing objects.

Figure 3.2 shows the basic idea of how the icon has been constructed. The *X*- and *Y*-axes are displayed in their correct orientations while the *Z*-axis is pointing outwards towards the user. The **W** on the icon indicates that the user is working with the world coordinate system. The origin is at the point (0,0,0) and is positioned at the lower left corner of the screen – as it is in 2D. The status bar displays the three coordinates of any point on the screen, but these figures can be misleading, especially when viewing in 3D. The origin point can be positioned to suit the model being created – more on this later.

The point P2 (30,40,50) is thus defined as 30 units in the positive *X*-direction, 40 units in the positive *Y*-direction and 50 units in the positive *Z*-direction. Similarly the point P3 (−40,−50,−30) has 40 units in the negative *X*-direction, 50 units in the negative *Y*-direction and 30 units in the negative *Z*-direction.

In the previous chapter, all the extruded models were created with the WCS.

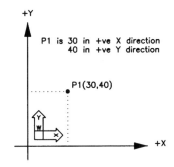

Figure 3.1 Coordinate entry with the WCS at the (0,0) origin.

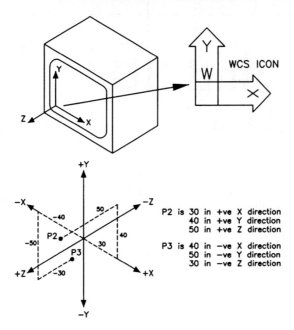

Figure 3.2 3D coodinate input.

The UCS

The UCS is probably the most important concept in 3D modelling and all users must be fully conversant with it. The user coordinate system allows the operator to:

a) move the origin to any point (or object) on the screen
b) align the UCS icon with existing objects
c) align the UCS icon to suit any 'plane' on a model
d) rotate the icon about the *X*-, *Y*- and *Z*-axes
e) save UCS 'positions'
f) recall previously saved UCS settings.

The appearance of the UCS alters depending on:

a) its orientation, i.e. how it is 'attached' to objects
b) the viewpoint selected or entered.

UCS icon exercise

As an introduction to the UCS icon, the following exercise is given as a sequence of operations which the reader should complete. No drawing is involved and it should be noted that several of the commands will be new to some readers, all of which will be explained later. The object of the exercise is to make the reader aware of the 'versatility' of the UCS icon.

1 Open your 3DSTDA3 template file and refer to Fig. 3.3.

2 The icon displayed at the lower left corner of the screen has a W on it, indicating that it is the WCS icon – fig. (a). This is the 'normal' default icon.

3 Menu bar with **View–Display–UCS Icon** and ensure both the On and Origin options are active, i.e. a tick at each name.

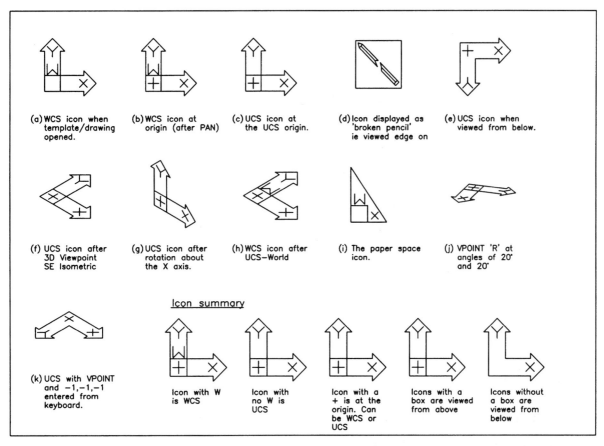

(a) WCS icon when template/drawing opened.

(b) WCS icon at origin (after PAN)

(c) UCS icon at the UCS origin.

(d) Icon displayed as 'broken pencil' ie viewed edge on

(e) UCS icon when viewed from below.

(f) UCS icon after 3D Viewpoint SE Isometric

(g) UCS icon after rotation about the X axis.

(h) WCS icon after UCS–World

(i) The paper space icon.

(j) VPOINT 'R' at angles of 20° and 20°

Icon summary

(k) UCS with VPOINT and −1,−1,−1 entered from keyboard.

Icon with W is WCS

Icon with no W is UCS

Icon with a + is at the origin. Can be WCS or UCS

Icons with a box are viewed from above

Icons without a box are viewed from below

Figure 3.3 WCS and UCS icon exercise.

4 At the command line enter **PAN <R>** and:
a) pan the screen upwards and to the right
b) right-click and pick Exit.

5 The icon will be displayed as fig. (b) with a + sign added at the 'box'. This + indicates that the icon is positioned at the origin.

6 With snap on, move the cursor onto the icon + and observe the status bar – the co-ordinates should be 0.00, 0.00, 0.00.

7 Pick the Undo icon from the Object Property toolbar to restore the screen to its original display.

8 Menu bar with **Tools–UCS–Origin** and:
prompt   Origin point<0,0,0>
enter   **100,100 <R>**
and   the icon moves to the entered point and is displayed as fig. (c). It has no W indicating that it is a UCS icon and has a + indicating it is at the origin.

9 Move the cursor onto the + and observe the coordinates in the status bar. They should display 0.00, 0.00, 0.00.

10 Menu bar with **Tools–UCS–X Axis Rotate** and:
prompt   Rotation angle about X axis<0.0>
enter   **90 <R>**
and   icon displayed as fig. (d). This is the AutoCAD 'broken pencil' icon indicating that we are looking at it 'edge-on'.

11 At the command line enter **UCS <R>** and:
   prompt   Origin/Zaxis/3 point/...
   enter   **X <R>** – the rotate about *X*-axis option
   prompt   Rotation angle about X axis
   enter   **90 <R>**
   and   icon displayed as fig. (e) and is being viewed from below – there is no 'box'.
                The + is still displayed indicating the UCS icon is still at the origin.

12 Menu bar with Tools–UCS–X Axis Rotate and enter 180 as the rotation angle. The icon will be displayed as fig. (c).

13 Menu bar with **View–3D Viewpoint–SE Isometric** and the icon will be displayed in 3D as fig. (f). It still has a + and is therefore still at the origin.

14 At the command line enter **UCS <R>** and:
   a) UCS options: enter **X <R>**
   b) rotation angle: enter **90 <R>**
   c) icon displayed as fig. (g).

15 Undo the UCS X rotation with U <R> or pick the Undo icon.

16 Menu bar with **Tools–UCS–World** and the icon will be displayed as fig. (h). The icon is still in 3D but is the world icon (W) and is still at the origin (+).

17 Menu bar with **View–3D Viewpoint–Plan View–World UCS** and the icon should be as the original fig. (a) – may need to pan?

18 Double-left click on the word MODEL in the status bar and the icon will be displayed as fig. (i). This is the paper space icon which will be discussed in a later chapter.

   At present undo this paper space effect with U <R> to restore the icon as fig. (a).

19 Enter/select the following sequences:
   a) Tools–UCS–Origin and enter **100,100** – fig. (c)
   b) enter **VPOINT** then R with angles of 20 and 20 – fig. (j)
   c) enter **VPOINT** the –1,–1,–1 to give the icon as fig. (k)
   d) enter **VPOINT** then R with angles of 0 and 90 – fig. (c)
   e) enter **VPOINT** then W – fig. (a).

20 This completes the icon exercise. There is no need to save.

21 *Note*: we could have used the UCS toolbar with icons during this exercise, but at our level I think that the menu bar and command line selections give the user a 'better understanding' of that is actually happening. You can investigate the UCS toolbar for yourself.

22 *Icon summary.*
   Figure 3.3 displays a summary of the various icons which can be displayed on the screen. These are:
   a) icon with a W is a WCS
   b) icon with no W is a UCS
   c) icon with a + is at the origin
   d) icon with a 'box' is viewed from above
   e) icon with no 'box' is viewed from below.

Orientation of the UCS

The completed exercise has demonstrated that the UCS icon can be moved to any point on the screen and rotated about the three axes (we only used the *X*-axis rotation, but the procedure is the same for the *Y*- and *Z*-axes). It is thus important for the user to be able to determine the correct orientation of the icon, i.e. how the *X*-, *Y*- and *Z*-axes are configured in relation to each other.

The axes orientation is determined by the **right-hand rule** and is demonstrated in Fig. 3.4. The knuckle of the right hand is at the origin and the position of the thumb, index finger and second finger determine the direction of the positive *X*-, *Y*- and *Z*-axes, respectively.

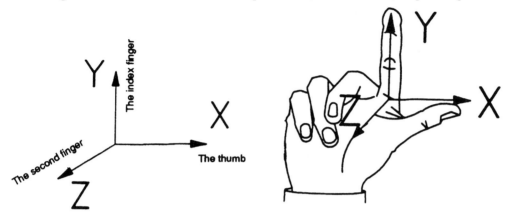

Figure 3.4 The right-hand rule.

Three-dimensional coordinate input

Coordinate input is generally required at some time during the creation of a 3D model. With 3D draughting there are three types of coordinates available, each having both absolute and relative entry modes. The three coordinate types are:

Type	Format	Absolute	Relative
Cartesian	x dist, y dist, z dist	100,150,120	@300, −100, −50
Cylindrical	dist<angle, Z dist	150 < 55,120	@75 < −15, −120
Spherical	dist<angle 1<angle 2	80 < 30 < 50	@120 < −10 < 75

To investigate the different types of coordinate input we will draw some objects on the screen. We will also investigate the effect of the icon position on the coordinate entries.

1   Open your 3DSTDA3 template file and refer to Fig. 3.5.

2   Menu bar with **View–Display–UCS Icon** and:
   a) on – tick
   b) origin – tick.

   These selections ensure that the icon is displayed on the screen and is 'set' to the origin point.

3   Menu bar with **View–3D Viewpoint–SE Isometric** to display the screen in 3D.

4   Menu bar with **View–Zoom–Scale** and:
   prompt   Enter scale factor and enter: **1 <R>**.

5   The WCS icon should be positioned at the left vertex of the black border – point A.

6   Make three new layers – L1, L2, L3 with continuous linetype and colour numbers 30, 72, 240, respectively.

WCS entry

1 With layer L1 current, use the LINE icon and draw:

From point	**0,0,0 <R>**		
To point	**150,100,80 <R>**	absolute	line 1W
To point	**@50,80,90 <R>**	relative absolute	line 2W
To point	**@100<30,80 <R>**	relative cylindrical	line 3W
To point	**@120<40<20 <R>**	relative spherical	line 4W
To point	**right-click.**		

2 Menu bar with **View–Zoom–All** to 'see' the lines.

3 Draw a circle, centre: 0,0,0 with radius: 50.

4 Add the following item of text:
   a) start point: 40,40,0
   b) height: 10 with 0 rotation
   c) item: AutoCAD WCS.

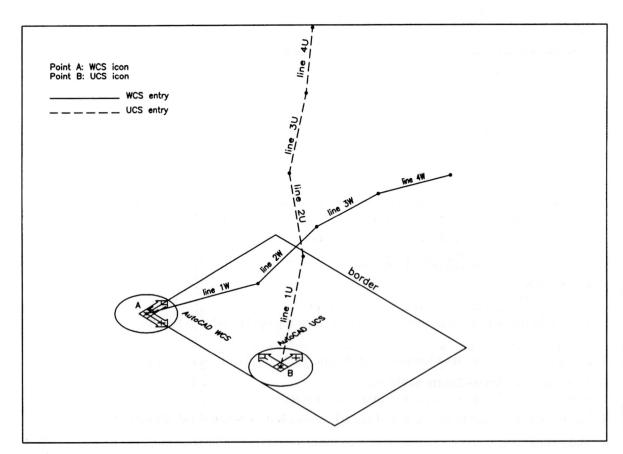

Figure 3.5 Coordinate entry exercise.

UCS entry

1 Menu bar with **Tools–UCS–Origin** and:
   prompt Origin point
   enter   **250,50,0 <R>**.

2 Menu bar with **Tools–UCS–Z Axis Rotate** and:
   prompt Rotation angle about Z axis
   enter   **90 <R>**.

3 The icon should be positioned and orientated at point B.

4 With layer L2 current, use the LINE icon and draw:

From point	**0,0,0 <R>**		
To point	**150,100,80 <R>**	absolute	line 1U
To point	**@50,80,90 <R>**	relative absolute	line 2U
To point	**@100<30,80 <R>**	relative cylindrical	line 3U
To point	**@120<40<20 <R>**	relative spherical	line 4U
To point	**right-click**.		

5 Draw a circle, centred on 0,0,0 with a 50 radius.

6 Add the text item:
   a) start point: 40,40,0
   b) height: 10 with 0 rotation
   c) item: AutoCAD UCS.

WCS entry with UCS icon

1 Make layer L3 current.

2 With LINE icon draw:

From point	***0,0,0 <R>**
To point	***150,100,80 <R>**
To point	**@*50,80,90 <R>**
To point	**@*100<30,80 <R>**
To point	**@*120<40<20 <R> the right-click**.

3 The drawn lines will be identical to the WCS entry lines.

Task

1 Save the coordinate exercise if required.

2 With **File–New** recall your 3DSTDA3 template file.

3 Menu bar with **View–Display–UCS icon** and ensure:
   a) on – tick
   b) origin – tick.

4 Menu bar with **View–3D Viewpoint–SE Isometric**.

5 Menu bar with **View–Zoom–Scale** and enter a factor of **1**.

6 The WCS icon should be positioned at left vertex of the border.

7 Save this layout as the **3DSTDA3.dwt** template file and:
   a) replace the existing template file
   b) do not alter the Template Description.

8 This will allow our template file to opened in 3D with the icon 'set' to the origin position.

Summary

1 There are two coordinate systems:
   a) the world coordinate system – WCS
   b) the user coordinate system – UCS.

2 Each system has its own icon.

3 The WCS is a fixed system, the origin being at 0,0,0.

4 The WCS icon is 'standard' and does not alter in appearance. The WCS icon is denoted with the letter W.

5 The UCS system allows the user to define the origin, either as a point on the screen or referenced to an existing object.

6 The UCS icon alters in appearance depending on the viewpoint.

7 The UCS icon can be rotated about the three axes.

8 The UCS current position can be saved and recalled.

9 3D coordinate input can be:
   a) cartesian, e.g. 10,20,30
   b) cylindrical, e.g. 10<20,30
   c) spherical, e.g. 10<20<30.

10 Both absolute and relative modes of input are possible with the three 'types' of coordinates, e.g.
   a) absolute cylindrical: 100<200,50
   b) relative cylindrical: @100<200,50.

11 3D coordinate input can be relative to the current UCS position or to the WCS, e.g.
   a) 100,200,150 for UCS entry
   b) *100,200,150 for WCS entry.

12 It is recommended that **3D coordinate input is relative to the current UCS position**.

Chapter 4

Creating a 3D wire-frame model

In this chapter we will create a 3D wire-frame model and use it to:
a) investigate how the UCS can be set and saved
b) add 'objects' and text to the model surfaces
c) modify the existing model.

Getting started

1 Open your 3DSTDA3 template file to display:
   a) a 3D viewpoint with the black border displayed
   b) the WCS icon at the left vertex of the border.

2 Ensure layer MODEL is current and refer to Fig. 4.1.

3 Display the Draw, Modify and Objects Snap toolbars.

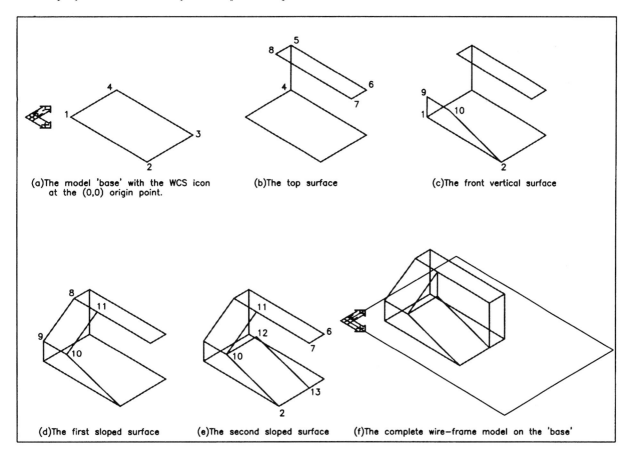

(a)The model 'base' with the WCS icon at the (0,0) origin point.

(b)The top surface

(c)The front vertical surface

(d)The first sloped surface

(e)The second sloped surface

(f)The complete wire–frame model on the 'base'

Figure 4.1 Construction of the wire-frame model 3DWFM.

Creating the wire-frame model
The base

1 With the LINE icon draw:

From point	**50,50 <R>**	pt1
To point	**@200,0 <R>**	pt2
To point	**@0,120 <R>**	pt3
To point	**@–200,0 <R>**	pt4
To point	**close** – fig. (a).	

The top surface

2 With the LINE icon draw:

From point	**Intersection icon of pt4**	
To point	**@0,0,100 <R>**	pt5
To point	**@200,0,0 <R>**	pt6
To point	**@0,–40,0 <R>**	pt7
To point	**@–200,0,0 <R>**	pt8
To point	**@0,40,0 <R>**	pt5
To point	right-click.	

3 Menu bar with **View–Zoom–Scale** and enter a scale factor of **0.9** to 'see' the complete construction – fig. (b).

The front vertical surface

4 LINE icon and draw:

From point	**Intersection icon of pt1**	
To point	**@0,0,45 <R>**	pt9
To point	**@60,0,0 <R>**	pt10
To point	**Intersection icon of pt2**	
To point	right-click – fig. (c).	

The first sloped surface

5 With the LINE icon draw:

From point	**Intersection of pt9**
To point	**Intersection of pt8** then right-click.

6 LINE icon again:

From point	**Intersection of pt10**	
To point	**Perpendicular to line 78**	pt11
To point	right-click – fig. (d).	

The second sloped surface

7 Select the LINE icon and draw:

From point	**Intersection of pt10**	
To point	**@0,80,0 <R>**	pt12
To point	**Perpendicular to line 23**	pt13
To point	right-click – fig. (e).	

Completing the model

8 The model requires three lines to be added, so with the LINE icon draw:
   a) from pt3 to pt6
   b) from pt7 to pt13
   c) from pt11 to pt12.

9 The completed model is displayed in fig. (f) on 'its base', i.e. the black border.

10 At this stage save the model as **R14MOD\3DWFM.dwg**.

11 *Note.* The model has been created using 3D coordinate input with the WCS, i.e. no attempt has been made to use the UCS. This is a perfectly valid method of creating wire-frame models, but difficulty can be experienced if objects and text have to be added to the various 'surfaces' of the model when the coordinates need to be calculated. Using the UCS usually overcomes this type of problem.

Moving around with the UCS

To obtain a better understanding of the UCS and how it is used with 3D models, we will use the created wire-frame model to add some objects and text. The sequence is quite long but it is important that you persevere and complete the exercise. Both menu bar and keyboard entry methods will be used to activate the UCS command.

1 Open the wire-frame model **R14MOD\3DWFM** or continue from the previous exercise. This model has the WCS icon at the black border origin point – the left vertex.

2 Refer to Fig. 4.2.

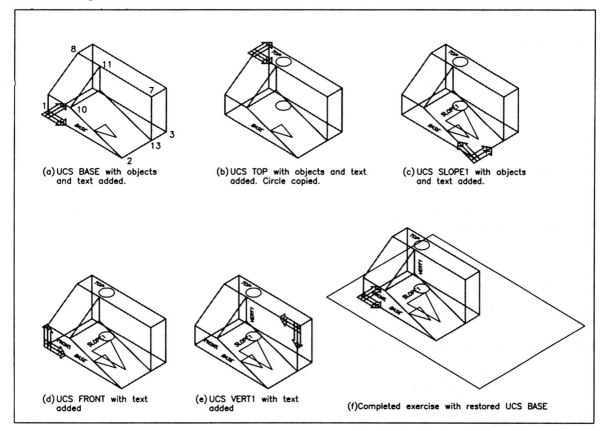

(a) UCS BASE with objects and text added.

(b) UCS TOP with objects and text added. Circle copied.

(c) UCS SLOPE1 with objects and text added.

(d) UCS FRONT with text added

(e) UCS VERT1 with text added

(f) Completed exercise with restored UCS BASE

Figure 4.2 Investigating the UCS and adding objects and text.

3 Menu bar with **View–Zoom–Scale** and enter a scale factor of 0.8 to allow us to 'see' the UCS movements.

4 Menu bar with **Tools–UCS–Origin** and:
   prompt   Origin point<0,0,0>
   respond   **Intersection icon and pick pt1**
   and   *a*) icon 'moves' to selected point – fig. (a)
                b) it is a UCS icon – no W
                c) it is at the origin – the +
   Note: if the icon does not move to the selected point, menu bar with **View–Display–UCS Icon** and pick Origin (i.e. tick).

5 Now that the icon has been repositioned at point 1, we want to save this 'position' for future recall, so menu bar with **Tools–UCS–Save** and:
   prompt   ?/Desired UCS name
   enter   **BASE <R>**.

6 Make layer OBJECTS current and use the LINE icon to draw:
   From   100,25,0
   To   @0,30,0
   To   145,40,0
   To   close.

7 Make layer TEXT current and menu bar with **Draw–Text–Single Line Text** and:
   a) start point: 60,10,0
   b) height: 8 and 0 rotation
   c) text item: BASE – fig. (a).

8 At the command line enter **UCS <R>** and:
   prompt   Origin/Zaxis/3point... i.e. the UCS options
   enter   **O <R>** – the origin option
   prompt   Origin point<0,0,0>
   respond   **Intersection icon and pick pt8**
   and   icon 'jumps' to the selected point – fig. (b).

9 At the command line enter **UCS <R>** and:
   prompt   UCS options
   enter   **S <R>** – the save option
   prompt   ?/Desired UCS name
   enter   **TOP <R>**

10 With layer OBJECTS current draw a circle with centre: 60,20 and radius: 15.

11 With layer TEXT current, add text using:
   a) start point: 20,25
   b) height: 8 with 0 rotation
   c) text item: TOP.

12 Using the COPY icon:
   a) select objects: pick the circle then right-click
   b) base point: Center icon and pick the circle
   c) second point: enter @0,0,–100 <R> – fig. (b)
   d) why these coordinates?

13 Menu bar with **Tools–UCS–3Point** and:
   prompt   Origin point<0,0,0>
   respond   **Intersection icon and pick pt2**
   prompt   Point on positive portion of the X-axis

respond   **Intersection icon and pick pt3**
prompt   Point on positive-Y portion of the UCS XY plane
respond   **Intersection icon and pick pt10**.

14   The UCS icon will move to point 2 and be 'aligned' on the sloped surface – fig. (c).
     Note: The 3-point option of the UCS command is 'asking the user' for three points to
         define the UCS icon orientation, these being:
         first prompt   the origin point
         second prompt   the direction of the *X*-axis
         third prompt   the direction of the *Y*-axis.

15   Save this UCS position with the menu bar selection **Tools–UCS–Save** and:
     prompt   ?/Desired UCS name
     enter   **SLOPE1 <R>**.   *DRG SAVED AS P25a*

16   With layer OBJECTS current use the LINE icon to draw:
     From   15,100,0
     To   @50,0,0
     To   40,30,0
     To   close.

17   With layer TEXT current, add a text item using:
     a) start point: centred on ~~40,110~~ *10,125*
     b) height: 8 with 0 rotation
     c) item: SLOPE1 – fig. (c).

18   Menu bar with **Tools–UCS–Restore** and:
     prompt   ?/Name of UCS to restore
     enter   **BASE <R>**
     and   icon restored to the base point as fig. (a).

19   Menu bar with **Tools–UCS–X Axis Rotate** and:
     prompt   Rotation angle about X axis
     enter   **90 <R>**
     and   Icon displayed as fig. (d).

20   At command line enter **UCS <R>** then **S <R>** for the save option and **FRONT <R>**
     as the UCS name.

21   With layer TEXT current add an item of text with:
     a) start point: ~~25,20~~ *20,15*
     b) height: 8 with 0 rotation
     c) text: FRONT – fig. (d).

22   At the command line enter **UCS <R>** and:
     prompt   UCS options
     enter   **3 <R>** – the three point option
     prompt   Origin point
     respond   **Intersection icon and pick pt7**
     prompt   X axis direction
     respond   **Intersection and pick pt11**
     prompt   Y axis direction
     respond   **Intersection icon and pick pt13**.

23   The UCS icon will be aligned as fig. (e).

24   Save this UCS position as VERT1 – easy? (UCS-S-VERT1).

25 With layer TEXT current add a text item with:
   a) start point: 120,50
   b) height: 8
   c) rotation: −90
   d) text: VERT1 – fig. (e).

26 Restore UCS BASE and the model will be displayed as fig. (f).

27 Save the drawing at this stage as ~~R14MOD\3DWFM~~ updating the original saved model.   *P26a*

Modifying the wire-frame model

To further investigate the UCS we will modify the wire-frame model, so:

1 3DWFM still on the screen? – if not open it.

2 Layer MODEL current with UCS BASE – fig. (a).

3 Refer to Fig. 4.3.

4 Select the CHAMFER icon from the Modify toolbar and:
   a) set both chamfer distances to 30
   b) chamfer lines 7–11 and 7–13
   c) chamfer lines 5–6 and 6–3.

   Now add two lines to complete the 'chamfered corner' and erase the unwanted original corner line – fig. (b).

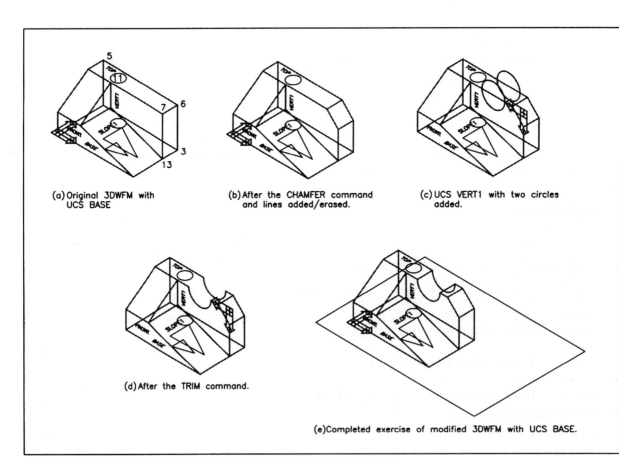

(a) Original 3DWFM with UCS BASE

(b) After the CHAMFER command and lines added/erased.

(c) UCS VERT1 with two circles added.

(d) After the TRIM command.

(e) Completed exercise of modified 3DWFM with UCS BASE.

Figure 4.3 Modifying the 3DWFM model.

5 Restore UCS VERT1 and note its position – fig. (c).

6 Draw two circles:
   a) centre at 80,0,0 with radius 30
   b) centre at 80,0,–40 with radius 30 – fig. (c).

7 Using the TRIM icon from the Modify toolbar:
   a) trim the two circles 'above' the model
   b) trim the two lines 'between' the circles – fig. (d).

8 Draw in the two lines on the top surface and restore UCS BASE.

9 The modified model is now complete – fig. (e).

10 Save the model as **R14MOD\3DWFM** updating the existing drawing.   27a

Note

1 The user should realize that the UCS is an important concept with 3D modelling. Indeed 3D modelling would be very difficult (if not impossible) without it.

2 I have continually used the word 'surface' when referring to the model. All wire-frame models are 'hollow' and do not have any surfaces as we know the word, but I hope that you understand what is meant with top surface, sloped surface, etc.

Task

The wire-frame model has 11 'flat surfaces' and one 'curved surface'. We have set and saved UCS positions for five of these surfaces – BASE, TOP, SLOPE1, FRONT and VERT1. You now have to set and save the other six flat UCS positions, i.e. one for each surface and add an appropriate text item to that surface.

My suggestions for the UCS name and text item are LEFT, RIGHT, REAR, SOLPE2, SLOPE3 and VERT2 but you can have any names of your choice.

Figure 4.4 displays the complete wire-frame model with text added to every surface (except the curved surface). When complete, remember to save as **R14MOD\3DWFM**   27e
as it will be used in other chapters.

Summary

1 Wire-frame models are created by coordinate input and by referencing existing objects.

2 Both the WCS and UCS entry modes can be used, but I would recommend:
   a) use the WCS to create the basic model outline
   b) use the UCS to modify and add items to the model.

3 It is strongly recommended that a UCS be set and saved for every surface on a wire-frame model.

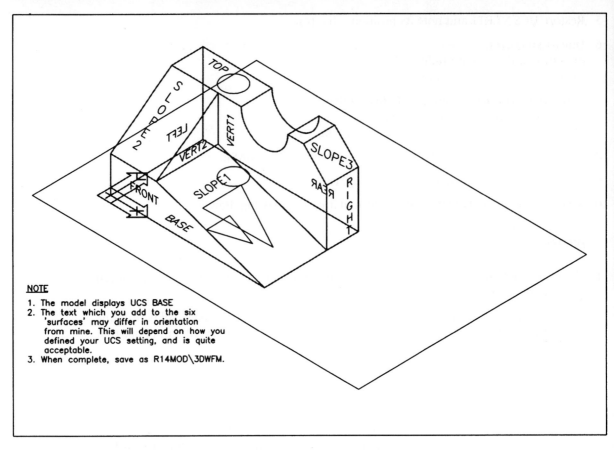

NOTE
1. The model displays UCS BASE
2. The text which you add to the six 'surfaces' may differ in orientation from mine. This will depend on how you defined your UCS setting, and is quite acceptable.
3. When complete, save as R14MOD\3DWFM.

Figure 4.4 The complete wire-frame model with text added to every 'surface'.

Assignments

Creating wire-frame models is important as it allows the user to:
a) use 3D coordinate entry with the WCS and/or the UCS
b) set and save different UCS positions
c) become familiar with the concept of 3D modelling.
I have included two 3D wire-frame models which have to be created. The suggested approach is:

1 Open your 3DSTDA3 template file.

2 Complete the model with layer MODEL current, starting at some convenient point, e.g. 50,50,0. Use WCS entry and add one surface at a time.

3 When the model is complete, set and save a UCS for every surface.

4 Add appropriate text.

5 Save each **DRAWING** as **R14MOD\ACT_2**, etc.

6 *Note*:
  a) **do not attempt to add dimensions**
  b) do not attempt to display the two models on 'one screen' – you will soon be able to achieve this for yourself.

Activity 2: Guide block

This is a relatively simple model to create. There are nine 'surfaces' which require text, therefore nine UCS positions.

Activity 3: Special slip block

Another fairly simple model to create with 10 surfaces requiring text to be added.

Chapter 5

The UCS

The UCS is one of the basic 3D draughting 'tools' and it has several commands associated with it. In this chapter we will investigate:
a) the UCS options
b) the UCS toolbar
c) the named UCS dialogue box
d) the UCS preset dialogue box
e) the PLAN command
f) the UCSFOLLOW variable.

Getting started

1 Open your R14MOD\3DWFM model from the previous chapter. This model has several blue objects with several saved UCS positions.

2 Restore the UCS BASE – probably is current?

3 Layer MODEL current and freeze layer TEXT. Refer to Fig. 5.1.

4 In the exercise which follows, the UCS command will be activated by command line entry but use the menu bar if you prefer.

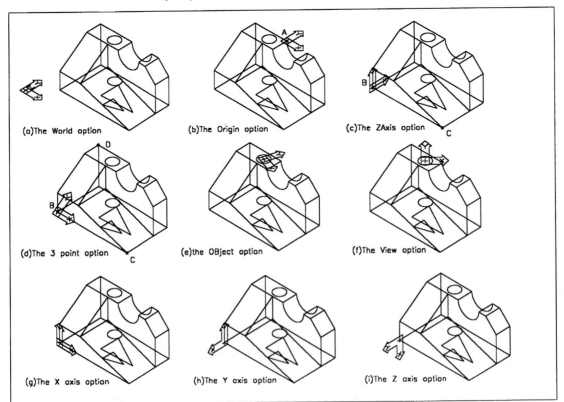

Figure 5.1 The UCS option exercise.

The UCS options

The UCS command has 14 options and can be activate from the menu bar with **Tools–UCS** or by entering **UCS** at the command line. The 14 options are:

Origin/Zaxis/3point/Object/View/X/Y/Z/Prev/Restore/Save/Del/?/<World>

World: this option restores the WCS setting irrespective of the UCS position. It is the default AutoCAD setting. At the command line enter **UCS <R>** then **W <R>** to display the WCS icon to 'left of model' – fig. (a).

Origin: used to set a new origin point. The user specifies this new origin point:
a) by picking any point on the screen
b) by coordinate entry
c) by referencing existing objects.
When used, the UCS icon is positioned at the selected point if the UCS Icon is set to Origin. This option has been used in previous exercises. At the command line enter **UCS <R>** then **O <R>** and:
prompt   Origin point<0,0,0>
respond   **Intersection icon and pick ptA** – fig. (b).

ZAxis: defines the UCS position relative to the *Z*-axis, the user specifying:
a) the origin point
b) any point on the *Z*-axis.
At the command line enter **UCS <R>** then **ZA <R>** and:
prompt   Origin point
respond   **Intersection icon and pick ptB**
prompt   Point on positive portion of Z-axis
respond   **Intersection icon and pick ptC** – fig. (c)
The icon will be aligned with:
a) the *X*-axis along the shorter base left edge
b) the *Y*-axis along the front left vertical edge
c) the *Z*-axis along the line BC.

3point: defines the UCS orientation by specifying three point:
a) the actual origin point
b) a point on the positive *X*-axis
c) a point on the positive *Y*-axis.
At the command line enter **UCS <R>** then **3 <R>** and:
prompt   Origin point
respond   **Intersection of ptB**
prompt   Point on positive portion of the X-axis
respond   **Intersection of ptC**
prompt   Point on positive-Y portion of the UCS XY plane
respond   **Intersection of ptD** – fig. (d).
This is a very useful option especially if the icon is to be aligned on sloped surfaces. It is probably my preferred method of setting the UCS.

Object: aligns the icon to an object, e.g. a line, circle, polyline, item of text, dimension, block, etc.
At the command line enter **UCS <R>** then **OB <R>** and:
prompt   Select object to align UCS
respond   **pick any point on circle on top surface**
The icon is aligned as fig. (e) with:
a) the origin at the circle centre point
b) the positive *X* axis pointing towards the circumference of the circle at the selected point.

View: aligns the UCS so that the *XY* plane is always perpendicular to the view plane. At the command line enter **UCS <R>** the **V <R>**.
The UCS icon will be displayed as fig(f) and is similar to the traditional 2D icon?
This is a useful UCS option as it allows 2D text to be added to a 3D drawing – try it for yourself.

X/Y/Z: allows the UCS to be rotated about the entered axis by an amount specified by the user.
  1 Restore UCS BASE
  2 At the command line enter **UCS <R>** then **X <R>** and:
    prompt Rotation angle about the X axis
    enter   **90 <R>** – fig. (g)
  3 At the command line enter **UCS <R>** then **Y <R>** and:
    prompt Rotation angle about the Y axis
    enter   **–90 <R>** – fig. (h)
  4 At the command line enter **UCS <R>** then **Z <R>** and:
    prompt Rotation angle about the Z axis
    enter   **–90 <R>** – fig. (i).

Prev: restores the previously 'set' UCS position and can be used to restore the last 10 UCS positions.

At the command line enter **UCS <R>** then **P <R>** continually until the WCS icon is returned as fig(a). This should be possible if you have not made any mistakes in the above exercises.

Restore: allows the user to restore a previously saved UCS position but the names of the saved UCS's must be remembered.
At the command line enter **UCS <R>** then **R <R>** and:
prompt   ?/Name of UCS to restore
enter   **one of your saved UCS names**.

Save: allows the user to save a UCS position for future recall. It should be used every time a new UCS has been defined. The option is activated from the command line with **UCS <R>** then **S <R>** and the user can enter a name with up to 31 characters.

Del: Entering **UCS <R>** then **D <R>** prompts for the UCS name to be deleted. The default is none. Use with care!

?:   the query option which will list all saved UCS positions At the command line enter **UCS <R>** then **? <R>** and:
prompt   UCS name(s) to list<*>
enter   *** <R>**
prompt   AutoCAD Text Window with a list of saved coordinate systems, i.e. names, origin points, *X*-, *Y*-, *Z*-axes details.
respond   cancel the window.

The UCS toolbar

All the UCS options have so far been activated by keyboard entry (**UCS <R>**) or from the menu bar (**View–Tools–UCS**). The only reason for this was that I think it is easier for the user to understand what option is being used. The UCS options can also be activated from the UCS toolbar – Fig. 5.2.

The UCS command has 14 options, all of which can be activated from the command line or from the menu bar. The UCS toolbar has no icon selection for the options Restore, Save, Delete or List (?), although these can easily be activated by selecting the UCS icon which displays the keyboard options. Two additional icons available in the toolbar are the Named UCS and the Preset UCS, both of which will now be discussed.

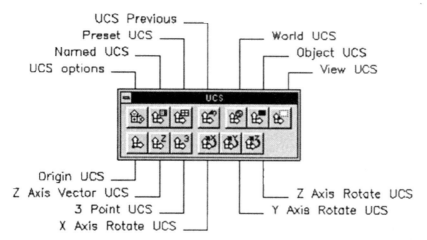

Figure 5.2 The UCS toolbar.

The UCS Control dialogue box (Named UCS)

Restoring a previously saved UCS is relatively simple as it involves:
a) keyboard: entering UCS–R–name
b) menu bar: Tools–UCS–Restore–name.

The problem with both these methods is that the user must remember the names of all the saved UCS positions. The list option (?) of the UCS command will allow the user to 'see' all the saved UCS names, but using the UCS Control dialogue box is easier.

1   Ensure R14MOD\3DWFM is displayed with UCS BASE current.

2   Menu bar with **Tools–UCS–Named UCS** and:
prompt   UCS Control dialogue box
with   1. list of all saved UCS names
                    2. BASE current
respond   1. pick **FRONT**
                    2. pick **Current** – Fig. 5.3
                    3. pick **OK**.

3   The UCS icon will be displayed in the FRONT setting.

4   The UCS Control dialogue box can also be activated with the NAMED UCS icon from the UCS toolbar.

5   Use the Named UCS icon to restore UCS TOP.

6   The UCS Control dialogue box is very useful as it allows:
a) the user to 'see' all the saved UCS names
b) the current UCS to be set, i.e. restored, and is easier than the two previous methods
c) the delete and list options
d) the ability to rename a UCS.

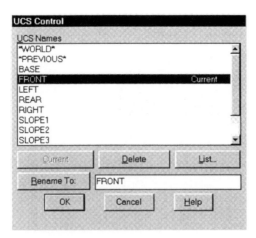

Figure 5.3   The UCS Control dialogue box.

The UCS Orientation dialogue box (Preset UCS)

AutoCAD has several preset UCS settings which can be activated from a dialogue box with:

a) menu bar **Tools–UCS–Preset UCS**

b) Preset UCS icon from the UCS toolbar.

1 Still with the 3DWFM on the screen with UCS TOP.

2 Select the Preset UCS icon from the UCS toolbar and:
  prompt   UCS Orientation dialogue box
  with   1 Relative to Current UCS active (black dot)
             2 TOP highlighted (in black)
  respond 1 pick **FRONT** – Fig. 5.4
             2 pick **OK**.

3 The UCS icon will be displayed in a FRONT orientation relative to the UCS TOP setting.

4 Select the Named UCS icon and restore UCS VERT1, i.e. make it current.

5 Select the Preset UCS icon, pick FRONT the OK. The UCS icon will be displayed in a FRONT orientation relative to UCS VERT1.

6 *Note*: the UCS Orientation dialogue must be used with caution. I prefer to set the UCS using the options.

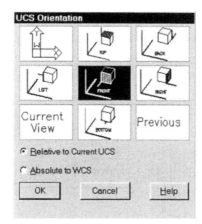

Figure 5.4 The UCS Orientation dialogue box.

Plan

Plan is a command which displays any model perpendicular to the *XY*-plane of the current UCS position.

1 Restore UCS BASE and refer to Fig. 5.5.

2 At the command line enter **PLAN <R>** and:
prompt   `<Current UCS>/Ucs/World`
enter   **<R>**, i.e. accept the Current UCS default.

3 The screen will display the model as a plan view – fig. (a). This view is perpendicular to the current UCS setting (BASE) and is really a 'top' view in orthogonal terms.

4 Restore UCS FRONT.

5 Menu bar with **View–3D Viewpoint–Plan View–Current UCS** and the model will be displayed as fig. (b). This is a plan view to the current UCS FRONT and is a 'front' view in orthogonal terms.

6 Menu bar with **View–3D Viewpoint–Plan View–Named UCS** and:
prompt   `?/Name of UCS`
enter   **SLOPE1 <R>**.

7 The model will be displayed as a plan to the UCS SLOPE1 setting as fig. (c).

8 At the command line enter **PLAN <R>** and:
prompt   `<Current UCS>/Ucs/World`
enter   **U <R>** – the UCS option
prompt   `?/Name of UCS`
enter   **VERT1 <R>**.

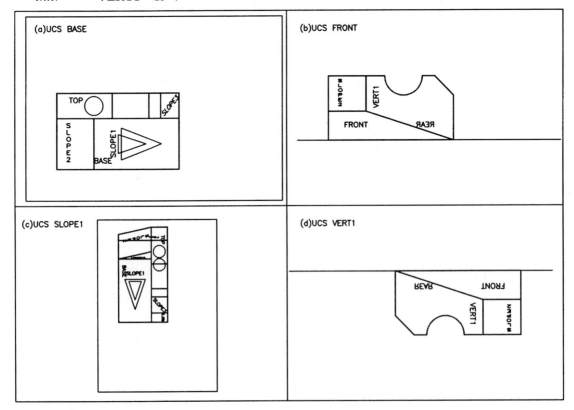

Figure 5.5 The PLAN command with 3DWFM.

9 The model display is as fig. (d), i.e. a plan view to the UCS setting VERT1. This display should be upside-down – why?

10 Finally restore UCS BASE and menu bar with View–3D Viewpoint–SE Isometric to return the original model display.

UCSFOLLOW

UCS FOLLOW is a system variable which controls the screen display of a model when the UCS position is altered. The variable can only have the values of 0 (default) or 1 and:
a) UCSFOLLOW 0: no effect on the display with UCS changes
b) UCSFOLLOW 1: generates a plan view when the UCS is altered.

1 Original 3D display with UCS BASE on the screen?

2 At the command line enter **UCSFOLLOW <R>** and:
prompt   New value for UCSFOLLOW <0>
enter   **1 <R>**.

3 Nothing has changed?

4 Restore UCS FRONT – plan view as Fig. 5.5(b).

5 Restore UCS SLOPE1 – plan view as fig. (c).

6 Restore UCS VERT1 – plan view as fig. (d).

7 Restore UCS BASE – plan view as fig. (a).

8 Set UCSFOLLOW to 0 and restore the original screen display with View-3D Viewpoint-SE Isometric.

9 This completes the exercises with the UCS.

Summary

1 The UCS is an essential factor with 3D modelling.

2 The UCS command has 14 options.

3 The orientation of the UCS icon is dependent on the option used.

4 The UCS toolbar offers fast option selection.

5 It is **strongly recommended** that the UCS icon and the UCS icon origin are ON when working in 3D. These can be activated with the menu bar sequence View–Display–UCS Icon.

6 The UCS Control dialogue box (Named UCS) is recommended for restoring previously saved UCS positions.

7 The UCS Orientation dialogue box (Preset UCS) should be used with caution.

8 PLAN is a command which displays the model perpendicular to the *XY* plane of the current UCS.

9 UCSFOLLOW is a system variable which can be set to give automatic plan views when the UCS is re-positioned.

Chapter 6

The modify commands with 3D models

All the modify commands are available for use with 3D models, but the results are dependent on the UCS position. We will investigate the COPY and ARRAY command with our 3D wire-frame model so:

1 Open R14MOD\3DWFM with UCS BASE.

2 Display the Modify and Object snap toolbars and refer to Fig. 6.1.

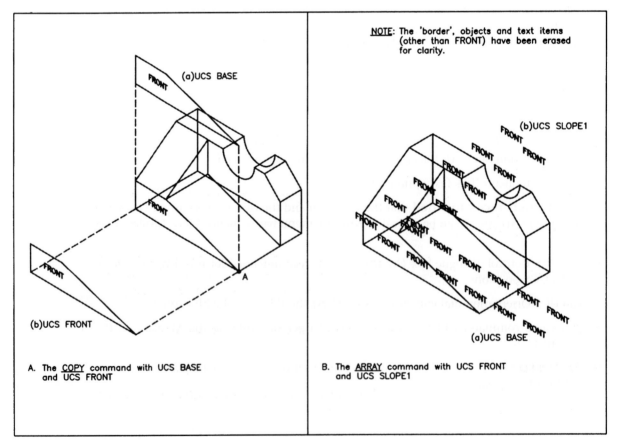

Figure 6.1 The COPY and ARRAY commands with 3DWFM.

The COPY command

1   Select the COPY icon from the Modify toolbar and:
   prompt   Select objects
   respond   **pick the 4 red lines and the green FRONT text item on the 'front vertical' surface then right-click**
   prompt   Base point
   respond   **Intersection icon and pick ptA**
   prompt   Second point of displacement
   enter   **@0,0,200 <R>** – fig. (a).

2   Restore UCS FRONT.

3   Select the COPY icon and:
   prompt   Select objects
   enter   **P <R><R>** – previous selection set option
   prompt   Base point and: **pick Intersection of ptA**
   prompt   Second point and enter: **@0,0,200 <R>** – fig. (b).

4   May need a Zoom–All?

5   Undo the copy effects if required.

The ARRAY command

1   Restore UCS BASE.

2   Select the ARRAY icon from the Modify toolbar and:
   prompt   Select objects
   respond   **pick the FRONT text item then right-click**
   prompt   Rectangular or Polar and enter: **R <R>**
   prompt   Number of rows and enter: **2 <R>**
   prompt   Number of columns and enter: **6 <R>**
   prompt   Row distance and enter: **40 <R>**
   prompt   Column distance and enter: **60 <R>**.

3   The text item is arrayed in a 2 × 6 rectangular matrix – fig. (a).

4   Restore UCS SLOPE1.

5   Rectangular array the FRONT text item using the same entries as step 2 – fig. (b).

Other modify commands

All of the modify commands are available for use with 3D models, but the final result is dependent on the UCS position. The only requirement for the user is to ensure that the icon is positioned on the 'plane' for the modification. Certain commands (e.g. Trim, Extend) will give the following message prompt:

View is not plan to UCS. Command results may not be as expected.

Summary

The AutoCAD modify commands with 3D models have to be used with care. The result is UCS dependent.

Chapter 7

Dimensioning in 3D

There are no special commands to add dimensions in 3D. Dimensioning is a 2D concept, the user adding the dimensions to the *XY* plane of the current UCS setting. This means that the orientation of the complete 'dimension object' will depend on the UCS position. The user should be aware of:

a) AutoCAD's automatic dimensioning facility

b) linear dimensioning will be horizontal or vertical, depending on where the dimension line is located in relation to the object being dimensioned.

We will demonstrate how dimensions can be added to 3D models with two examples. The first will be the 3D wire-frame model 3DWFM, and the second will use AutoCAD's stored 3D objects.

Example 1

1 Open R14MOD\3DWFM and display the Dimension, Object Snap and other toolbars to suit.

2 Freeze layer TEXT and make layer DIM current.

3 *Note*: the standard sheet created for the template file had a created dimension style – 3DSTD. You may want to 'alter' the Overall Scale (Geometry selection) to a higher value than the 1 default.

4 Ensure UCS BASE is current and refer to Fig. 7.1.

5 *a*) select the LINEAR dimension icon and:

    prompt   First extension line origin
    respond   **Intersection icon and pick pt1**
    prompt   Second extension line origin
    respond   **Intersection icon and pick pt2**
    prompt   Dimension line location
    respond   **pick to suit**

  b) repeat the LINEAR dimension selection and dimension line 23, positioning the dimension line to suit

  c) select the DIAMETER icon and:

    prompt   Select arc or circle
    respond   **pick the circle on the TOP surface**
    prompt   Dimension line location
    respond   **pick to suit** – interesting result?

  d) the three added dimensions will be displayed as fig. (a).

6 Erase the added dimensions and restore UCS FRONT.

7 Using the dimension icons:

  a) linear dimension lines 12 and 14

  b) align dimension line 56

  c) try and add a diameter dimension to the top circle

  d) dimensions displayed as fig. (b).

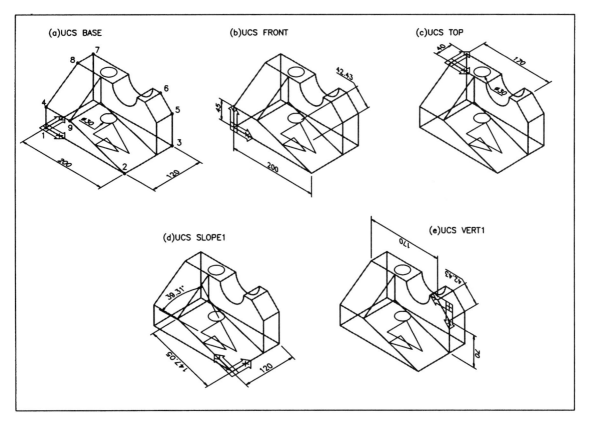

Figure 7.1 Dimension exercise with 3DWFM.

8 Erase these added dimensions and restore UCS TOP and:
   a) linear dimension line 67 and line 78
   b) diameter dimension the circle on the top
   c) result as fig. (c).

9 Restore UCS SLOPE1, erase the previous dimensions and:
   a) linear dimension line 23 and line 29
   b) angular dimension a vertex of the blue triangle 'on the slope'
   c) the three dimensions will be displayed as fig. (d).

10 With UCS VERT1 current, erase the dimensions from SLOPE1 and:
   a) linear dimension line 67 and line 35
   b) align dimension line 56
   c) interesting result as fig. (e) – why?

11 *Note*: This exercise should demonstrate to the user that:
   a) adding dimensions to a 3D model is **very UCS dependent**
   b) there are no special 3D dimension commands
   c) the actual orientation of added dimensions depends on the UCS
   d) dimensions are added to the *XY* plane of the current UCS.

12 *Task*: *a*) erase any dimensions still displayed
       b) with layer DIM current (should be) refer to Fig. 7.2 and add the given dimensions to the model
       c) some of the existing saved UCS positions will be used
       d) you may have to set a new UCS position for the continuous 80,40 and the 70 dimensions?
       e) when complete save if required, but not as 3DWFM.

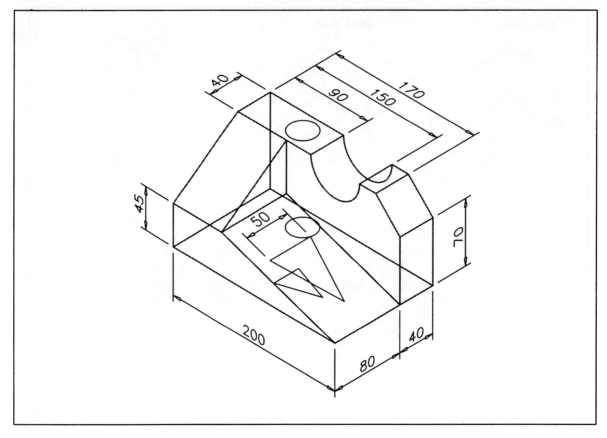

Figure 7.2 Required dimensions to be added to 3DWFM.

Example 2

1 Menu bar with **File–New** and:
   a) pick **Use a Template**
   b) pick **3DSTDA3.dwt** – your 3D template file
   c) pick **OK**.

2 The screen will display a black border:
   a) in SE Isometric viewpoint
   b) layer MODEL current
   c) WCS icon at left vertex of border.

3 Display the Surfaces toolbar.

4 Menu bar with **Draw–Surfaces–3D Surfaces** and:
   prompt   3D Objects dialogue box
   respond **pick Box3D then OK**
   prompt   Corner of box and enter: **50,50,0 <R>**
   prompt   Length and enter: **120 <R>**
   prompt   Cube/<Width> and enter: **80 <R>**
   prompt   Height and enter: **60 <R>**
   prompt   Rotation about Z axis and enter: **0 <R>**.

5 Select the WEDGE icon from the surface toolbar and:
prompt Corner of wedge and enter: **200,150,0 <R>**
prompt Length and enter: **100 <R>**
prompt Width and enter: **75 <R>**
prompt Height and enter: **65 <R>**
prompt Rotation about Z axis and enter: **0 <R>**.

6 Using the 3-point UCS option:
   a) origin at point 1
   b) *x* axis along line 12
   c) *y* axis along line 13
   d) save UCS position as POS1
   e) add two linear dimensions with this UCS.

7 Use the 3-point UCS option again with:
   a) origin at point a
   b) *x* axis along line ab
   c) *y* axis along line ac
   d) save as POS2
   e) add two linear dimensions with this UCS.

8 *Task*. Add the three other dimensions, i.e. one to the box and two to the wedge. Some UCS manipulation is required but you should manage this without any problems.

9 Save if required, but the drawing will not be used again.

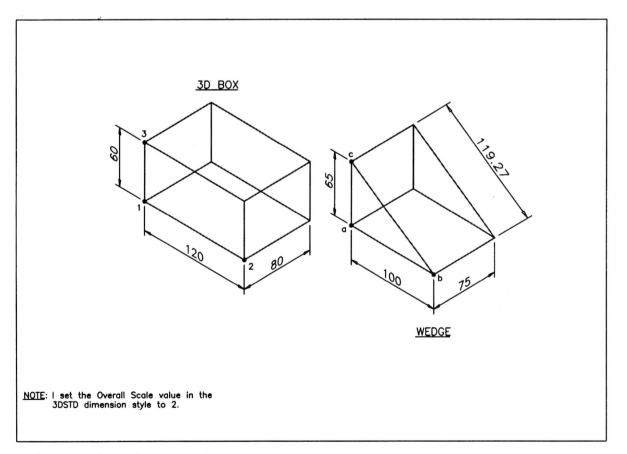

NOTE: I set the Overall Scale value in the 3DSTD dimension style to 2.

Figure 7.3 Adding dimensions to 3D objects.

Summary

1 There are no special 3D dimension commands.

2 Dimensioning is a 2D concept and dimensioning a 3D model involves adding the dimensions to the *XY* plane of the required UCS setting.

If the UCS is not positioned correctly, dimensions can have the 'wrong orientation' in relation to the object being dimensioned.

Assignment

Most CAD users will know that the Egyptians were famous for the pyramids and other landmarks. They may not be aware that one of the least known of their builders was Macfaramus who created several unusual structures. It is two of his designs which are included as wire-frame models, the user having to create and dimension both models. As these models will be used in other activities, ensure that they are completed (even if the dimensions are not added).

Activity 4: Shaped block

Macfaramus decided to deviate from the traditional cuboid shaped block used to build the pyramids and designed his own unusual shaped block. You have to:
a) create a wire-frame model of the block using the sizes given
b) set and save five UCS positions using the information given
c) add the given dimensions
d) save the completed block as **R14MOD\SHBLOCK**.

Activity 5: Square-topped pyramid

The square-topped pyramid designed by Macfaramus was never built but his design is still considered unique. You have to:
a) create a wire-frame model of the pyramid using the sizes given
b) set and save several UCS positions, these being on the following surfaces:
  1. the base and top – obvious
  2. four sloped surfaces – names and orientation given
  3. four vertical surfaces – should give you no problems. Use your own names, e.g. V1,V2,V3 and V4
c) add the given dimensions and any other?
d) save the completed model as **R14MOD\PYRAMID**.

Chapter 8

Hatching in 3D

There are no special 3D hatch commands. Hatching (like dimensioning) is a 2D concept, the hatch pattern being added to the *XY* plane of the current UCS. Two examples will be used to demonstrate adding hatching to 3D models.

Example 1

1 Open your 3DSTDA3.dwt template file.

2 Display the Draw, Modify and Object Snap toolbars.

3 With layer MODEL current, refer to Fig. 8.1 and draw the four 'mutually perpendicular planes' using the LINE icon with:

	Plane 1234	*Plane 1564*	*Plane 3764*	*Plane 7896*
From	30,30,0	30,30,0	130,130,0	130,130,100
To	@100,0,0	@0,0,100	@0,0,100	@0,100,0
To	@0,100,0	@0,100,0	@−100,0,0	@−100,0,0
To	@−100,0,0	@0,0,−100	right-click	@0,−100,0
To	close	right-click		right-click.

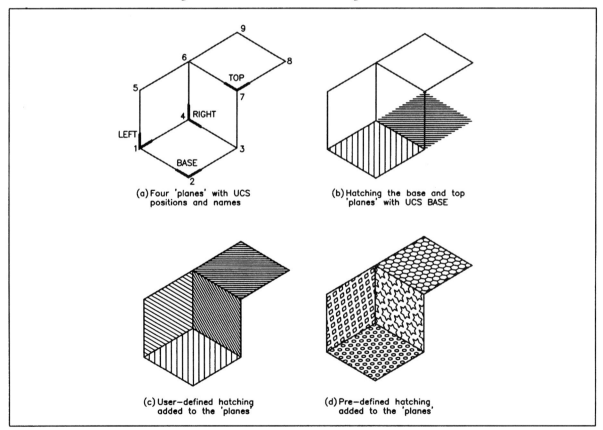

(a) Four 'planes' with UCS positions and names

(b) Hatching the base and top 'planes' with UCS BASE

(c) User-defined hatching added to the 'planes'

(d) Pre-defined hatching added to the 'planes'

Figure 8.1 3D hatch example 1.

4 Erase the black border and pan the drawing to suit.

5 Set and save the four UCS positions as fig. (a).

6 Restore UCS BASE and make SECT the current layer.

7 Select the HATCH icon from the Draw toolbar and using the Boundary Hatch dialogue box:
   a) pick User-defined pattern type
   b) set angle to 45 and spacing to 8
   c) use the **Pick Points<** option and:
      i) select a point within the 1234 plane then right-click
      ii) Preview Hatch–Continue–Apply.

8 Repeat the HATCH icon selection and:
   a) using the Pick Points option pick a point within the 6789 plane
      prompt   Boundary Definition Error dialogue box – Fig. 8.2
      respond   **pick OK** then right-click
   b) using the Select Objects option pick the four lines of the 6789 plane then right-click and:
      i) set angle to –45 and spacing to 5
      ii) Preview–Continue–Apply.

9 The result of the two hatch operations is displayed in fig. (b). Plane 1234 has the correct hatching, but plane 6789 has none, the hatching having been added to the plane of UCS BASE.

10 Use the HATCH icon and try and add hatching to the vertical planes 1564 and 3467. Possible?

11 Erase the 'wrong' hatching and restore UCS TOP.

12 Hatch the top plane (6789) using the HATCH icon with:
   a) pick points option
   b) angle –45 and spacing 5.

13 Add hatching to the two vertical planes remembering to restore UCS LEFT and UCS RIGHT – fig. (c).

14 *Task*.
   a) Erase the added hatching
   b) add the following pre-defined hatch patterns using the information given:

UCS	Pattern	Scale	Angle
BASE	HEX	1	–10
TOP	HONEY	2	0
LEFT	SQUARE	2	10
RIGHT	STARS	2	0

   c) the result is fig. (d)
   d) this completes example 1, which does not have to be saved.

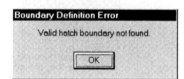

Figure 8.2   Boundary Definition Error dialogue box.

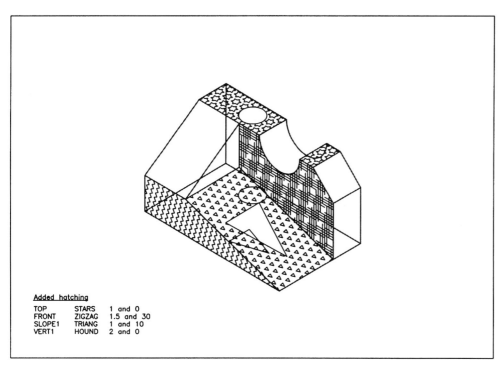

Added hatching

TOP	STARS	1 and 0
FRONT	ZIGZAG	1.5 and 30
SLOPE1	TRIANG	1 and 10
VERT1	HOUND	2 and 0

Figure 8.3 Hatch exercise with 3DWFM.

Example 2

1 Open the drawing R14MOD\3DWFM and refer to Fig. 8.3.

2 Make layer SECT current and freeze layers TEXT and DIM.

3 Restore UCS TOP.

4 Select the HATCH icon and:
   a) pattern type: STARS
   b) scale: 1 and angle: 0
   c) pick points: pick points in the **two** top surface planes then right-click
   d) preview–continue–apply.

5 Restore UCS FRONT and with the HATCH icon:
   a) pattern type: ZIGZAG
   b) scale: 1.5 and angle: 30
   c) select objects: pick the four lines of front surface then right-click
   d) preview–continue–apply.

6 Repeat the HATCH icon selection and add the following hatch patterns:

UCS	Pattern	Scale, angle	Type
SLOPE1	TRIANG	1,10	pick points
VERT1	HOUND	2,0	pick points.

7 Menu bar with **View–Hide** and the model is as before. Hatching a wire-frame model does not produce a hide effect.

8 Save your completed hatched model as R14MOD\3DWFM.

9 *Note*. In the two hatch exercises we used the one layer (SECT) for all hatching. It is sometimes desirable to have a different layer for each current UCS that is to be used for hatching. This is a procedure which is recommended.

Summary

1 Hatching is a 2D concept.

2 Hatching a 3D model requires the UCS to be set to the 'plane' which is to be hatched.

3 *a*) both user-defined and pre-defined hatch patterns can be used

   b) both the select objects and pick points methods are permitted.

4 It is recommended that a hatch layer is made for each 'surface' which is to be hatched.

5 Pre-R14 users will know that hatching can use disc memory. AutoCAD claim that Release 14:
   a) has 'solved' the scale factor problem
   b) does not use 'bytes and bytes' of memory
   c) as a point of interest, my 3DWFM had the following size:
      Figure 7.2: 3DWFM with text and dimensions – 42000
      Figure 8.3: 3DWFM with hatching added – 45000
   d) perhaps 'they' are now correct?

Assignments

Two hatch activities have been included, both using the created wire-frame models from the previous chapter, i.e. SHBLOCK and PYRAMID. In both activities:

1 Open the saved drawing.

2 Freeze layer DIM, or erase the dimensions.

3 Make new hatch layers for each UCS position – your decision!

4 Use the pick points method where possible.

Activity 6: The shaped block

Restore the appropriate UCS and add the following user-defined hatching:

UCS	Angle,Spacing
FRONT	45,8
RIGHT	–45,8
SLOPE1	45,8
SLOPE2	–45,8

When complete save the model as **R14MOD\SHBLOCK** – it will be used in later chapters.

Activitity 7: the pyramid

Using the correct UCS, add the following predefined hatch patterns:

UCS	Pattern	Scale,Angle
all sloped surfaces	BRICK	2,0
all vertical surfaces	BRSTONE	1,0
the top surface	EARTH	1.5,0

Save the complete hatched model as **R14MOD\PYRAMID**.

Note: my single drawing of Activities 6 and 7 are on the one A3 sheet of paper. With all hatching added, the total memory size of drawing was 81,516. This is quite small when considered to previous releases.

Chapter 9

Tiled viewports

The graphics screen can be divided into a number of separate viewing areas called **viewports** and each viewport can display any part of a drawing. Viewports are **interactive**, i.e. what is drawn in one viewport is automatically drawn in the others and the user can switch between viewports when creating a model. Viewport layouts (**configurations**) can be saved thus allowing different displays of the same model to be stored for future recall. Viewports are essential with 3D modelling as they allow different views of the model to be displayed on the screen simultaneously. When used with the VIEWPOINT command (next chapter) the user has a very powerful 3D draughting tool.

There are two types of viewport, displayed in Fig. 9.1, these being:

1 Tiled or fixed.

2 Untiled or floating.

The type of viewport which is displayed is controlled by the **TILEMODE** system variable and:
a) Tilemode 1: tiled viewports – cannot be moved (default setting)
b) Tilemode 0: untiled viewports – can be moved.

In this chapter we will only investigate TILED viewports and leave the untiled viewport discussion to a later chapter when we will investigate model and paper space.

The viewport command can be activated by keyboard entry or from the menu bar.

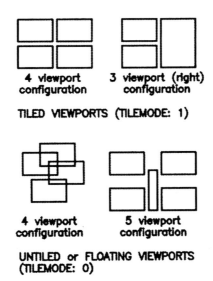

Figure 9.1 · Tiled and untiled viewports.

Example 1

1 Open your 3DWFM drawing of the wire-frame model on the black border. This exercise is rather long.

2 Make layer MODEL current and restore UCS BASE. Deactivate all toolbars.

3 Display the model without any text, dimensions or hatching. Erase or freeze layers – your choice.

4 At the command line enter **TILEMODE <R>** and:
prompt   New value for TILEMODE<1>
and   observe the 1 default then **ESC**.

5 The TILEMODE value of 1 indicates that only TILED viewports can be used. The same condition is also evident with:
   a) Status bar: word TILE in bold type
   b) Menu bar: View selection – Model Space (Tiled) is ON, i.e. tick.

6 Menu bar with **View–Tiled Viewports–Layout** and:
prompt   Tiled Viewport Layout dialogue box
respond   1. pick Three: Right – Fig. 9.2
              2. pick **OK**.

7 The drawing screen will:
   a) be divided into three separate 'areas' – one large at the right and two smaller to the left. The three viewports will 'fill the screen' – Fig. 9.3(a)
   b) display the same view of the model in the three viewports but at different 'sizes'.

8 Move the mouse about the screen and:
   a) the large viewport will display the cursor cross-hairs and is the **active** viewport, i.e. it is 'current'
   b) the other viewports will display an arrow and these viewports are **non-active**.

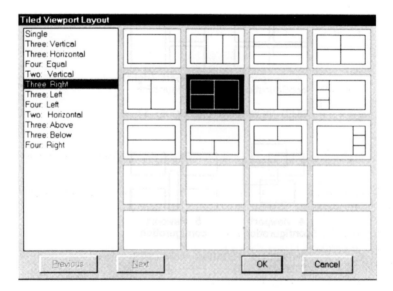

Figure 9.2 Tiled Viewport Layout dialogue box.

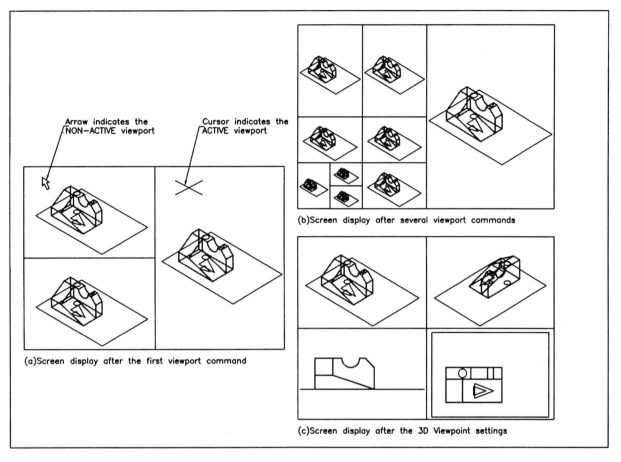

Figure 9.3 Viewport example 1.

9   Any viewport can be made active by:
   a) moving the mouse into the viewport area
   b) left-click
   c) try this for yourself a few times.

10 Menu bar with **View–Tiled Viewports–Save** and:
   prompt   `?/Name for new viewport configuration`
   enter   **CONF1 <R>**.

11 Make the upper left viewport active and select the menu bar sequence **View–Tiled Viewports–2 Viewports** and:
   prompt   `Horizontal/<Vertical>`
   enter   **V <R>**
   and   the top left viewport will be further divided into two equal vertical viewports, each displaying the model.

12 Make the lower left viewport active and menu bar with **View–Tiled–Viewport–4 Viewports** to display an additional four viewports of the model.

13 At the command line enter **VPORTS <R>** and:
   prompt   `Save/Restore...`
   enter   **S <R>** – the save option
   prompt   `?/Name for new viewport configuration`
   enter   **CONF2 <R>**.

14 With the lower left viewport active, enter **VPORTS <R>** at the command line and:
    prompt   Save/Restore...
    enter   **3 <R>** – the 3-viewport option
    prompt   Horizontal/Vertical/Above...
    enter   **L <R>** – the left option.

15 The lower left viewport will be further divided into another three viewport configuration with the large viewport to the left.

16 At this stage your screen should resemble Fig. 9.3(b).

17 Make the lower left viewport active and enter **VPORTS <R>** then **4 <R>** and the following message will be displayed at the prompt line: The current viewport is too small to divide.

18 Save the screen viewport configuration as CONF3 – easy?

19 *a*) Make the large right viewport active
    b) menu bar with View–Tiled Viewports–1 Viewport
    c) original screen display?
    d) Zoom–All needed?

20 Menu bar with **View–Tiled Viewports–4 Viewports** to 'fill the screen' with four viewports of the model.

21 Using the menu bar **View–3D Viewpoint** selection make each viewport current and set different viewpoints using the following information:
    Viewport   *3D Viewpoint*
    top left   SE Isometric
    top right   NE Isometric
    lower right   Plan-Current UCS
    lower left   Front.

22 The screen display should resemble Fig. 9.3(c).

23 Save the screen configuration as CONF4.

24 *Task*: return the screen to a single viewport configuration to display the original model.

25 Menu bar with **View–Tiled Viewports–Restore** and:
    prompt   ?/Name of viewport configuration to restore
    enter   **CONF1 <R>**
    and   screen displays the first saved configuration.

26 At the command line enter **VPORTS <R>** and:
    prompt   Save/Restore...
    enter   **R <R>** – the restore option
    prompt   ?/Name of viewport configuration to restore
    enter   **CONF2 <R>**.

27 Restore the other saved viewport configurations.

28 This completes the first exercise. If you want to save this exercise (with the viewport configurations) do **not** use the name **3DWFM**.

Example 2

The first exercise used an already created 3D model to investigate the viewport command and configurations. This exercise will create a new 3D wire-frame model interactively using a four viewport configuration with preset 3D viewpoints. This will allow the user to 'see' the model being created in all four viewports at the one time.

1   Open your **3DSTDA3.dwt** template file to display the black border in a 3D viewpoint with layer MODEL current. Refer to Fig. 9.4.

2   Check the status bar – TILE in bold type, i.e. tiled viewports.

3   Menu bar with **Tools–UCS–Origin** and enter **50,50,0** as the new origin point. Has icon moved? If not View–Display–UCS Icon–Origin.

4   Save the UCS position as BASE.

5   Set a four viewport configuration with **View–Tiled Viewports–4 Viewports**.

6   Making the appropriate viewport active, use the menu bar sequence **View–3D Viewpoint** to set the following viewpoints:
    Viewport   *3D Viewpoint*
    lower right   Plan-Current UCS
    lower left   SE Isometric
    upper left   Right
    upper right   Front.

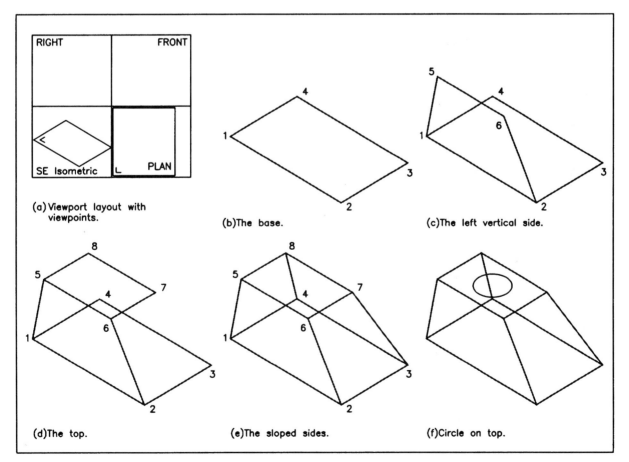

(a) Viewport layout with viewpoints.

(b) The base.

(c) The left vertical side.

(d) The top.

(e) The sloped sides.

(f) Circle on top.

Figure 9.4   Construction of viewport example 2.

7 In the upper left and upper right viewports, enter **ZOOM <R>** and:
   prompt   All/Center/Dynamic...
   enter   **1 <R>**.

8 The screen display should resemble fig. (a).

9 With the lower left viewport active, construct the model base using the LINE icon with:
From point	**0,0,0 <R>**	pt1
To point	**@200,0,0 <R>**	pt2
To point	**@0,120,0 <R>**	pt3
To point	**@200<180,0 <R>**	pt4
To point	**close** – fig. (b).	

10 Using the LINE command construct the left vertical side with:
From point	**Intersection icon of pt1**	
To point	**@20,0,100 <R>**	pt5
To point	**@120,0,0 <R>**	pt6
To point	**Intersection icon of pt2**	
To point	**right-click** – fig. (c).	

11 The top surface is created with the LINE command and:
From point	**Intersection icon of pt6**	
To point	**@0,80,0 <R>**	pt7
To point	**@–120,0,0 <R>**	pt8
To point	**Intersection icon of pt5**	
To point	**right-click** – fig. (d).	

12 Add the sloped sides with lines joining points 3–7 and 4–8 as fig. (e).

13 Make layer OBJECTS (blue) current and draw a circle with: centre: 80,40,100 and radius: 25 – fig. (f).

14 Menu bar with **Draw–Surfaces–3D Surfaces–Box3d** and:
prompt	Corner of box and enter: **80,30,0**
prompt	Length and enter: **50**
prompt	Width and enter: **40**
prompt	Height and enter: **30**
prompt	Rotation angle and enter: **20**.

15 *Task*.
   a) Make layer TEXT current
   b) Rotate UCS about X axis and save as FRONT
   c) Add the text item AutoCAD, centred on 80,50 with height 10 and rotation 0
   d) Set a 3 point UCS on the right sloped surface and save as SLOPE
   e) Add the text item Release 14, start point to suit with a height of 10 and rotation of 0.

16 The complete four viewport configuration display should be similar to Fig. 9.5.

17 Save the drawing as **R14MOD\TEST3D**.

18 This completes the two exercises on viewports.

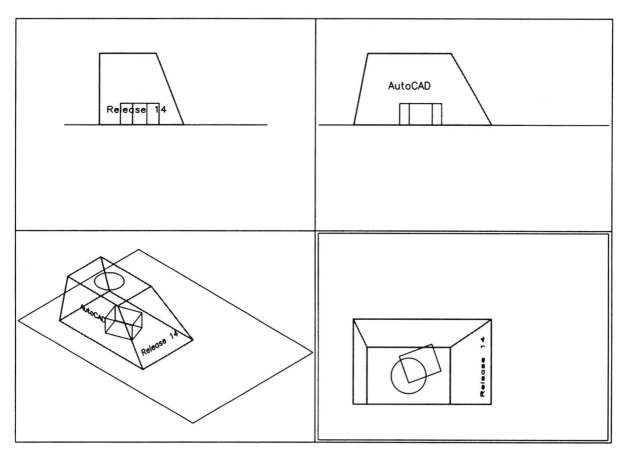

Figure 9.5 Completed viewport example 2 – TEST3D.

Note

1 The commands REGEN and REGENALL are useful with 3D modelling and:
   a) REGEN: refreshes the active viewport
   b) REGENALL: refreshes all viewports.

   Both commands can be activated from the command line or from the View menu bar selection.

2 The menu bar selection View–Tiled Viewports–? Viewports will divide the active viewport into sub-divisions until this is no longer possible – **viewport too small** message.

3 The menu bar selection View–Tiled Viewports–Layout... will display the layout dialogue box. Selecting any option will alter the complete screen layout and not just the active viewport.

Summary

1 Viewports allow multi-screen configurations.

2 There are two types of viewport – TILED and UNTILED.

3 The viewport type is controlled by the system variable TILEMODE and:
TILEMODE 1: tiled viewports (fixed)
TILEMODE 0: untiled viewports (movable) – more later.

4 Tiled viewports can have between 1 and 4 screen displays.

5 Multi-screen viewports are used with the viewpoint command – next chapter.

6 Multiple viewport layouts are extremely useful with 3D modelling.

Assignments

No specific activities for this chapter. Viewports activities will be assigned after the chapter on viewpoints.

Chapter 10

The viewpoint command

Viewpoint is the command which determines how the user 'looks' at a model and has been used in previous chapters. In this chapter we will investigate the command in detail using previously created models. When combined with viewports, the user has a very powerful draughting aid – multiple viewports displaying different views of a model.

The viewpoint command has four main options, these being:
a) the rotate option
b) the tripod (bull's eye and target) option
c) the vector option
d) the 3D viewpoint presets.

The actual command can be activated from the menu bar or by direct entry at the command line.

Viewpoint ROTATE option

This option requires two angles:
a) the angle from the *XY*-plane from the *X*-axis – the **view** direction
b) the angle from the *XY*-plane – the **inclination (tilt)**.

1 Open your R14MOD\3DWFM drawing and erase any dimensions and hatching. Leave the text items – they will act as a 'reference' as the model is viewed from different angles.

2 Layer MODEL current, UCS BASE and SE Isometric viewpoint.

3 Refer to Fig. 10.1, section A.

4 Menu bar with **View–3D Viewpoint–Rotate** and:

prompt	***Switching to WCS***
then	Enter angle in XY plane from X-axis
enter	**40 <R>**
prompt	Enter angle from XY plane
enter	**0 <R>**
prompt	***Returning to UCS***
then	Regenerating drawing
and	model displayed as fig. (a1), i.e. looking towards the right-rear side from a horizontal 'stand-point'.

5 At the command line enter **VPOINT <R>** and:

prompt	***Switching to WCS***
prompt	Rotate/<View point>...
enter	**R <R>** – the rotate option
prompt	Enter angle in XY plane... and enter: **90 <R>**
prompt	Enter angle from XY plane and enter: **0 <R>**
and	model displayed as fig. (a2).

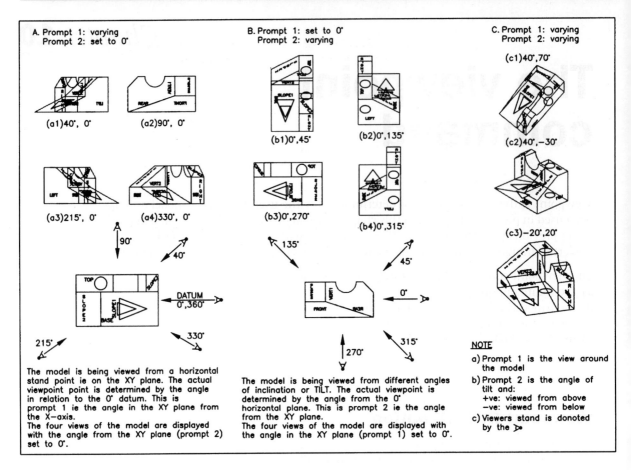

A. Prompt 1: varying
   Prompt 2: set to 0°

(a1)40°, 0°   (a2)90°, 0°

(a3)215°, 0°   (a4)330°, 0°

90°
40°

TOP
DATUM
0°,360°

215°   330°

The model is being viewed from a horizontal
stand point ie on the XY plane. The actual
viewpoint point is determined by the angle
in relation to the 0° datum. This is
prompt 1 ie the angle in the XY plane from
the X-axis.
The four views of the model are displayed
with the angle from the XY plane (prompt 2)
set to 0°.

B. Prompt 1: set to 0°
   Prompt 2: varying

(b1)0°,45°   (b2)0°,135°

(b3)0°,270°

(b4)0°,315°

135°   45°

0°

270°   315°

The model is being viewed from different angles
of inclination or TILT. The actual viewpoint is
determined by the angle from the 0°
horizontal plane. This is prompt 2 ie the angle
from the XY plane.
The four views of the model are displayed with
the angle in the XY plane (prompt 1) set to 0°.

C. Prompt 1: varying
   Prompt 2: varying

(c1)40°,70°

(c2)40°,-30°

(c3)-20°,20°

NOTE
a) Prompt 1 is the view around
   the model
b) Prompt 2 is the angle of
   tilt and:
   +ve: viewed from above
   -ve: viewed from below
c) Viewers stand is donoted
   by the ⏵

Figure 10.1 Viewpoint – the ROTATE option.

6 Repeat the viewpoint rotate command (menu bar or command line) and enter the following angle values at the prompts:

prompt 1	prompt 2	fig.
215	0	a3
330	0	a4.

7 Restore the original SE Isometric viewpoint and refer to Fig. 10.1 section B.

8 Use the viewpoint rotate command (menu bar or keyboard?) and enter the following angles at the prompts:

prompt 1	prompt 2	fig.
0	45	b1
0	135	b2
0	270 (–90)	b3
0	–45 (315)	b4.

9 Restore the SE Isometric viewpoint and refer to Fig. 10.1 section C. Activate the viewpoint rotate command and enter the following angles:

prompt 1	prompt 2	fig.
40	70	c1
40	–30	c2
–20	20	c3.

10 Restore the original viewpoint.

11 *Task*.

Restore some other UCS settings, e.g. SLOPE1, VERT1, etc. and repeat the viewpoint rotate command using some of the above angle entries. The model display should be unaffected by the UCS. Think about the prompt: ***Switching to the WCS***.

12 *Explanation of option*.

a) Prompt 1: angle in the *XY* plane from the *X*-axis. This is the viewer's standpoint **on the XY horizontal plane** looking towards the model, i.e. it is your view direction. If this angle is 0° you are looking at the model from the right side. If the angle is 270° you are looking onto the front of the model. The value of this angle can be between 0° and 360°. It can also be positive or negative and remember that 270° is the same as −90°.

b) Prompt 2: angle from the *XY* plane. This is the viewer's 'head inclination' looking at the model, i.e. it is the **angle of tilt**. A 0° value means that you are looking at the model horizontally and a 90° value is looking vertically down. The angle of tilt can vary between 0° and 360° and be positive or negative and:

positive tilt: looking down on the model
negative tilt: looking up at the model.

Viewpoint ROTATE using a dialogue box

1 3DWFM displayed at SE Isometric setting with UCS BASE?

2 Menu bar with **View–3D Viewpoint–Select...** and:
   prompt   Viewpoint Presets dialogue box – Fig. 10.2
   with   1. viewing angle: absolute to WCS
              2. angle from *X*-axis: 315 – left-hand 'clock'
              3. angle from *XY* plane: 35.3 – right-hand 'arc'.

3 This dialogue box allows:
   a) viewing angle to be absolute to WCS or relative to UCS
   b) angles to be set by selecting circle/arc position
   c) angles to be set by altering values at **From**: line
   d) plan views to be set.

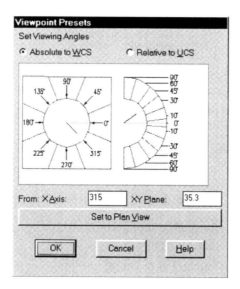

Figure 10.2   Viewpoint Presets dialogue box.

4 Respond to the dialogue box with:
   a) do not change to absolute to WCS
   b) change the *X*-axis angle from 315 to 150
   c) change the *XY* plane angle from 35.3 to 10
   d) pick **OK**
   e) the model will be displayed at the entered viewpoint.

5 Make UCS SLOPE1 current.

6 Menu bar with View–3D Viewpoint–Select and:
   a) restore Relative to UCS – black dot
   b) leave the two angle values as 150 and 10
   c) pick **OK**
   d) the model is displayed at the entered viewpoint but is differs from the step 4 display due to the UCS setting.

7 *Task*.
   a) Try some other entries from the Viewpoint Presets dialogue box using both selection methods, i.e. the clock/arc and altering the angles.
   b) Investigate the difference in the display with the Absolute to WCS and Relative to UCS selections.
   c) Restore UCS BASE and the SE Isometric viewpoint.

8 This completes the viewpoint rotate exercise. Do not save any changes.

Viewpoint TRIPOD option

This option allows the user to set 'infinite viewpoints' and is also called the bull's-eye and target method – for obvious reasons. We will demonstrate the command with a different model so:

1 Open the R14MOD\TEST3D model created during the viewport exercise and refer to Fig. 10.3.

2 Ensure UCS BASE is current and make the lower left viewport active, i.e. the 3D viewport.

3 Menu bar with **View–Tiled Viewports–Layout** and:
   prompt   Tiled Viewports Layout dialogue box
   respond   **pick Single then OK**
   and   screen displays a single viewport of the model in a 3D viewpoint and 'fills the screen'.

4 Menu bar with **View–3D Viewpoint–Tripod** and:
   prompt   1. model 'disappears'
               2. screen displays the *XYZ* tripod and the bull's-eye target
   and   axes and cross(+) move as mouse is moved
   respond   move the cross (+) into the circle quadrant indicated in fig. (a) and left-click
   and   model displayed at this viewpoint – viewed from above?

5 At the command line enter **VPOINT <R>** and:
   prompt   ***Switching to WCS***
   prompt   Rotate/<View point>...
   respond   **right-click**
   prompt   tripod and axes
   respond   move the cross (+) into the circle quadrant indicated in fig. (b) and left-click
   and   model displayed at this new viewpoint.

6 Repeat the tripod viewpoint option (menu bar or command line) and position the cross (+) in the quadrants indicated in Fig. 10.3, i.e. (a)–(d): within the inner circle and (e)–(h): within the outer circle.

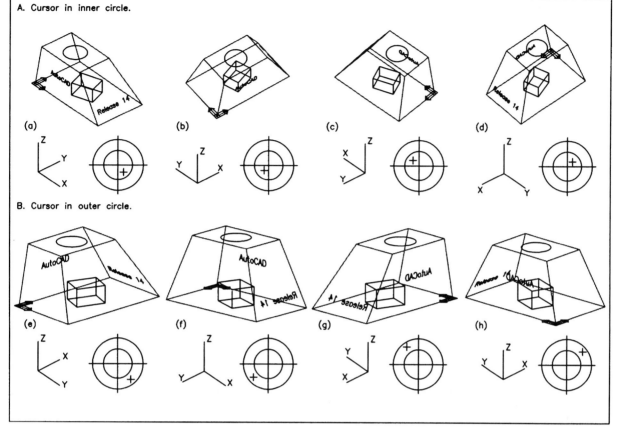

A. Cursor in inner circle.

(a)   (b)   (c)   (d)

B. Cursor in outer circle.

(e)   (f)   (g)   (h)

Figure 10.3 Viewpoint – the TRIPOD option.

7 *Task.*
When you are capable of using the tripod, try the following:
a) position the + at different points on the two axes and observe the resultant displays
b) position the + at different points on the circle circumferences and note the displays.

8 *Explanation of option.*
a) The 'bull's-eye' is in reality a representation of a glass globe and the model is located at the centre of the globe. The *XY*-plane is positioned at the equator. The north pole of the globe is at the circle centres and the two concentric circles represent the surface of the world, stretched out onto a flat plane with:
circle centre: the north pole
inner circle: the equator
outer circle: the south pole.

b) As the cross (+) is moved about the circles, the user is moving around the surfaces of the globe and:

Cross (+) position	View result
in inner circle	above equator, looking down on model
in outer circle	below equator, looking up at model
on inner circle	looking horizontally at model
below horizontal	viewing from the front
above horizontal	viewing from the rear.

9 This completes the tripod option exercise. Do not save any of these changes.

Viewpoint VECTOR option

1 Open R14MOD\3DWFM with UCS BASE and SE Isometric viewpoint. Erase any dimensions and hatching, but leave the text items – they will act as a 'reference' as the model viewpoint is altered.

2 Refer to Fig. 10.4.

3 Menu bar with **View–3D Viewpoint–Vector** and:
prompt ***Switching to WCS***
prompt Rotate/<View point><1.00,-1.00,1.00>
enter **0,0,1 <R>**.

4 The model will be displayed at this viewpoint. It is a top view as fig. (a) and 'fills the screen'.

5 At the command line enter **VPOINT <R>** and:
prompt ***Switching to WCS***
prompt Rotate/<View point><0.00,0.00,1.00>
enter **0,–1,0 <R>**
and model displayed as fig. (b) – a front view.

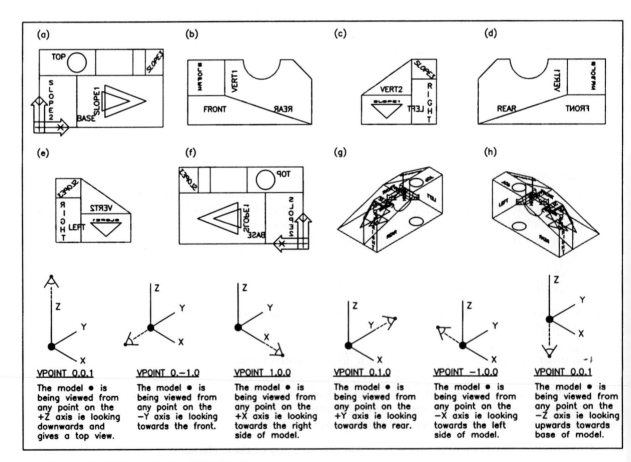

Figure 10.4 Viewpoint – the VECTOR option.

6 Repeat the viewpoint vector option (menu bar or command line) and enter the following coordinates at the prompt:

coords	view	fig.
1,0,0	from right	c
0,1,0	from rear	d
−1,0,0	from left	e
0,0,−1	from below	f
1,1,1	3D from above	g
−1,−1,−1	3D from below	h.

7 Restore the original SE Isometric viewpoint.

8 *Task.*
Try some vector entries for yourself then restore the original SE Isometric viewpoint.

9 *Explanation of option*
   a) The vector option allows the user to enter x,y,z-coordinates. These are the coordinates of the viewer's 'standpoint' looking at the model which is considered to be at the origin. Thus if you enter 0,0,1 you are 'standing' at the point 0,0,1 looking towards the origin. As this point is on the positive Z-axis you are looking down on the model, i.e. a top view.
   b) The actual numerical value of the vector entered does not matter, i.e. 0,0,1; 0,0,12; 0,0,99.99; 0,0,3456 are all the same viewpoint entries. I prefer to use the number 1, hence 0,0,1; −1,0,0, etc.
   c) Certain vector entries give the same display as rotate options and the following lists some of these similarities:

vector	rotate	view
0,0,1	90,90	top
0,−1,0	270,0	front
1,0,0	0,0	right
0,1,0	90,0	rear
−1,0,0	180,0	left
0,0,−1	90,−90	bottom
1,1,1	45,35	3D from above
−1,−1,−1	−135,−35	3D from below.

10 This completes the vector option. Do not save any changes.

The 3D viewpoint presets

The 3D viewpoint presets have been used in previous chapters and allow the user:
a) access to all the viewpoint options
b) four isometric viewpoints, e.g. SE Isometric, etc.
c) six 'traditional' views, e.g. top, right, etc.
d) plan views.

The preset option can be activated:
a) from the menu bar with View–3D Viewpoint
b) from the Viewpoint toolbar – does not give the plan selection.

It is the user's preference as to which method is used to activate the various viewpoint options.

The VIEW command

Different views of a model can be saved within the current drawing, thus allowing the operator to create a series of 'pictures'. These could be of the model being constructed, of a completed model at differing viewpoints, etc. These views (pictures) can be recalled at any time.

1 Still with 3DWFM on the screen with UCS BASE and SE Isometric viewpoint?

2 At the command line enter **VIEW <R>** and:
   prompt ?/Delete/Restore/Save...
   enter **S <R>** – the save option
   prompt View name to save
   enter **V1 <R>**.

3 Menu bar with View–3D Viewpoint and set to NW Isometric.

4 At the command line enter **VIEW <R>** and:
   prompt ?/Delete/Restore... and enter: **S <R>**
   prompt View name to save and enter: **V2 <R>**.

5 With View–3D Viewpoint set a Front view, the use the VIEW command to save the view as V3.

6 At the command line enter **VIEW <R>** and:
   prompt ?/Delete/Restore...
   enter **? <R>**
   prompt View(s) to list<*> and: **right-click**
   prompt AutoCAD Text Window with:
            Saved Views:
            View name   Space
            V1   M
            V2   M
            V3   M
   respond cancel the text window.

7 Menu bar with **View–Named Views** and:
   prompt View Control dialogue box as Fig. 10.5
   respond **pick V1–Restore–OK**
   and original view of model restored.

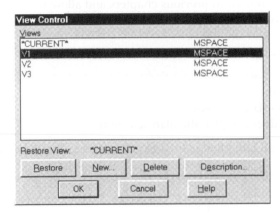

Figure 10.5 View Control dialogue box.

8 *Task*.
a) Create and save some other views
b) investigate both the Description and New options from the View Control dialogue box
c) restore saved views from the View Control dialogue box and from the command line.

9 When complete, restore the SE Isometric viewpoint. Save if required but not as 3DWFM.

10 *Note*.
Do not confuse the VIEW command with the View option of the UCS command. They are two entirely different concepts.

Summary

1 The viewpoint command allows models to be viewed from different 'standpoints'.

2 The command has several options – rotate, tripod, vector, presets.

3 The rotate option requires two angles:
a) the angle 'around' the model – the direction
b) the angle of inclination – the tilt.

4 The rotate option can be set from a dialogue box.

5 The tripod option allows unlimited viewpoints.

6 The vector option requires an x,y,z-coordinate entry.

7 The 3D Viewpoint presets are useful for 'set' viewpoints.

8 Viewpoints are generally set **absolute to the WCS** and the relative to the UCS option is **not recommended**.

9 All wire-frame models exhibit **ambiguity** when the viewpoint command is used, i.e. viewed from above or from below?

10 The VIEW command allows different views of a model to be saved in the current drawing for future recall. This is useful when the model is being displayed at various viewpoints.

Assignments

No specific activities for the viewpoint command.

Chapter 11

Centring viewports

When 3D models are displayed in multiple viewport configurations, three 'problems' can initially occur:

a) the model may 'fill the viewport'
b) the model may be displayed at different sizes in the viewports
c) the model views may not 'line up' between viewports.

These 'problems' are easily overcome by zooming each viewport about a specified centre point determined by the user. The user then decides on the 'scale effect' in the viewports. We will demonstrate the concept with two previously created models.

Example 1

1 Open 3DWFM of the hatched model and:
   a) erase any dimensions
   b) freeze layer TEXT
   c) leave the hatching displayed
   d) erase the black border
   e) Zoom–all and the model 'fills the screen'
   f) ensure UCS BASE is current.

2 Menu bar with **View–Tiled Viewports–4 Viewports** and the model is displayed in 3D in each viewport.

3 Set the following viewpoints in the named viewport:
viewport	*viewpoint*
top left	Right
top right	Front
lower right	Top
lower left	SE Isometric.

4 The model will be displayed at the viewpoints entered and will be of differing sizes in each viewport. The model needs to be centred about a specific point.

5 The model is basically of a cuboid shape with overall dimensions of $200 \times 120 \times 100$ and if UCS BASE is current, the model 'centre' is at the point 100,60,50.

6 Make the lower right viewport active.

7 Menu bar with **View–Zoom–Center** and:
   prompt   Center point and enter: **100,60,50 <R>**
   prompt   Magnification or Height<??> and enter: **150 <R>**.

8 The model will be 'centred' in the active viewport.

9 Repeat the zoom–center command in the other three viewports and:
   a) centre point: 100,60,50
   b) magnification: 150 – but 250 in the 3D viewport.

10 When the zoom-center command has been completed, the model will be 'neatly centred' in all viewports as Fig. 11.1.

11 Save as R14MOD\MV3DWFM.

Example 2

1　Open drawing TEST3D to display the four viewport configuration of the created model with text on two sloped surfaces.

2　Erase the black border and zoom–all in each viewport and the model will be displayed at different 'sizes'.

3　Ensure UCS BASE is current.

4　The model has overall sizes of $200 \times 120 \times 100$ and its 'centre' relative to UCS BASE is the point 100,60,50.

　　Note: It is pure coincidence that this centre point is the same as example 1.

5　In each viewport zoom–centre using:
　　a) centre point: 100,60,50
　　b) magnification: 175, but 200 in the 3D viewport.

6　The model will be centred in each viewport as Fig. 11.2.

7　Save this display as R14MOD\MVTEST3D.

8　This completes the centring exercises.

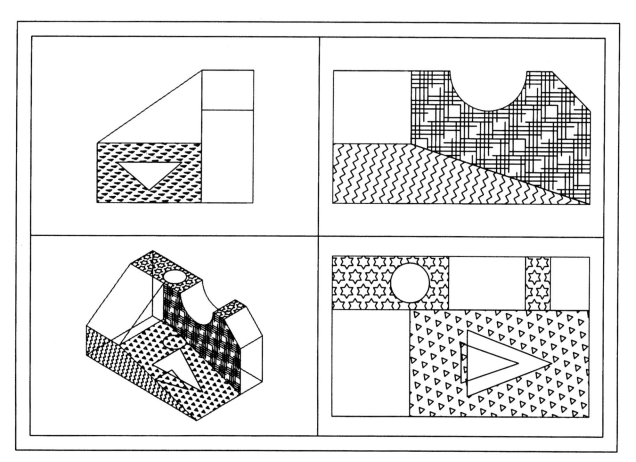

Figure 11.1 Centre viewport example 1 – 3DWFM.

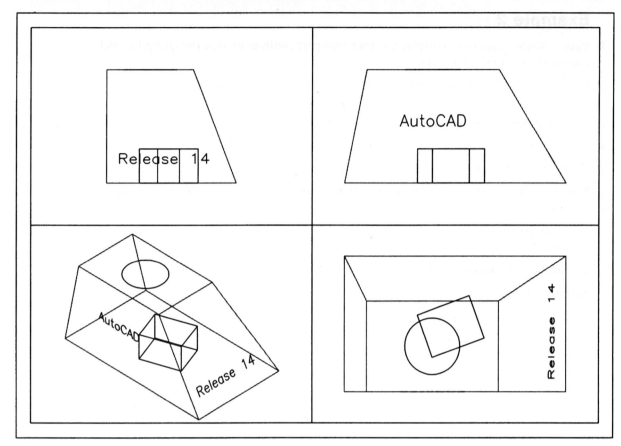

Figure 11.2 Centre viewport example 2 – 3DWFM.

Summary

1   Models can be centred in viewports using the Zoom–Center command, the user specifying:
    a) the centre point
    b) a magnification or height.

2   The centre point entered is **dependent on the UCS position**.

3   I like to zoom–centre relative to UCS BASE which is usually 'set' at a convenient base vertex. It is then easy for me to work out the model centre point in relation to the overall size of the model.

4   The magnification value entered is a 'scale' effect and is relative to the given default, e.g. if the default is <180> then:
    a) a value less than 180 will increase the model size
    b) a value greater than 180 will decrease the model size.

Assignments

Two activities have been included at this stage, both of which involve creating multiple viewports, setting viewpoints and centring the models. The models have already been created (hopefully) during the hatching activities.

Activity 8: The hatched shaped block of Macfaramus

1 Open the drawing R14MOD\SHBLOCK of the hatched block – Activity 6.

2 Create a four viewport configuration to display top, front, right and 3D viewpoints.

3 With UCS BASE current, zoom centre about the point 100,60,55 at 225 magnification in all viewports.

4 When complete, save as R14MOD\MVBLOCK.

Activity 9: The hatched pyramid of Macfaramus

1 Open R14MOD\PYRAMID from Activity 7.

2 Create a four viewport configuration to display top, front, right and 3D viewpoints.

3 With UCS BASE current, zoom centre about the point 100,100,95 at 275 magnification in all viewports.

4 When complete, save as R14MOD\MVPYR.

5 *Note*: in my activity drawing I have frozen certain hatch layers in specific viewports. You cannot (yet) do this?

Chapter 12

Model space and paper space – untiled viewports

AutoCAD Release 14 has multi-view capabilities which allow the user to layout, organize and plot multiple views of any 3D model. The multiple viewport concept has already been used in the creation of our wire-frame models, these viewports being TILED, i.e. fixed.

In this chapter we will investigate how to create UNTILED or FLOATING viewports which are used in exactly the same way that the tiled viewports were used. The creation of untiled viewports requires an understanding of the two drawing environments – model space and paper space.

Model space

This is the drawing environment that exists in any viewport and is the default. All models which have been created have been completed in model space. Model space is used for all draughting and design work and for setting 3D viewpoints. Multiple viewports are possible in model space but are TILED, i.e. they cannot be moved or altered in size – Fig. 12.1(a). While model space multi-views are useful, they have one major disadvantage – only the active viewport can be plotted, i.e. model space multiple viewports cannot be plotted on one sheet of paper.

Paper space

This is a drawing environment which is independent of model space. In paper space the user creates the drawing sheet, i.e. border, title box, etc. as well as arranging the multiple viewport layout. The viewports created in paper space are UNTILED, i.e. they can be positioned to suit, altered in size and additional viewports can be added to the layout – Fig. 12.1(b). In paper space the 3D viewpoint command is not valid although objects (particularly text) can be added to the sheet layout. The real advantage of working with paper space multiple viewports is that any viewport configuration can be plotted on the one sheet of paper.

Tilemode

The system variable which controls the 'type' of viewport to be created is **TILEMODE** and:
a) TILEMODE 1:   model space (FIXED) viewports
                      paper space is not available
b) TILEMODE 0:   paper space (FLOATING) viewports
                      model space is available.

Tiled (model space) viewports are always displayed as edge-to-edge and fill the screen like a tiled wall. Untiled (paper space) viewports can be positioned anywhere within the screen area with spaces between them if required. They can also be copied, moved, stretched, etc.

> **Both types of viewport cannot be used in one drawing.**

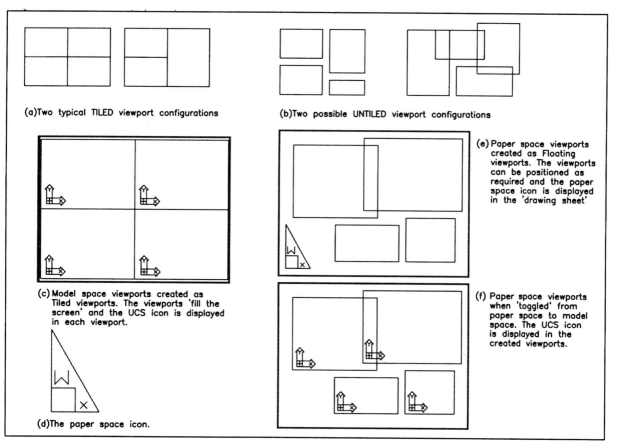

Figure 12.1 Model and paper space concepts.

Icons

When working in model space the normal WCS/UCS icon will be displayed in all viewports, orientated to the viewport viewpoint as Fig. 12.1(c). In paper space, the paper space icon – Fig. 12.1(d) is displayed.

When viewports are created in paper space, the paper space icon is displayed in the lower left corner of the screen as Fig. 12.1(e), but when model space 'is entered' the UCS icon is again displayed in all created viewports as Fig. 12.1(f).

Toggling between model and paper space

All AutoCAD users will be familiar with the toggle concept, e.g. toggling the grid on/off with the F7 key or from the Status bar with a double left-click on the word GRID. It is possible to toggle between model space and paper space but only if the TILEMODE system variable is set to 0. The toggle effect can be activated by:

1 *Command line*
   a) if in paper space, toggle to model space with **MS <R>**
   b) if in model space, toggle to paper space with **PS <R>**

2 *Status bar*
   a) double left-click on PAPER to toggle to model space
   b) double left-click on MODEL to toggle to paper space

3 *Menu bar*
  a) if in model space, select View–Paper Space
  b) if in paper space, select View–Model Space (Floating)

Preference: user decides, but I prefer the PS/MS entry.

Model/paper space example – untiled viewports

This example will use a model created in model space to demonstrate the paper space multiple viewport concept. The example is quite long but if you are unsure of paper space, persevere with it – it is important that you understand how to create and use paper space viewports.

1 Open drawing R14MOD\3DWFM and:
  a) erase the black border, hatching and any dimensions
  b) leave the text items – they will 'act as a reference'
  c) layer MODEL and UCS BASE current
  d) zoom–all and model 'fills the screen'
  e) refer to Fig. 12.2.

2 Make a new layer called VP, linetype continuous, colour to suit.

3 At the command line enter **PS <R>** and prompt:

    ** Command not allowed unless TILEMODE is set to 0 **

4 Menu bar with **View–Paper Space** and:
  a) blank screen returned
  b) paper space icon displayed
  c) where is the model?
  d) we have entered the paper space environment.

5 Make layer 0 current and draw a rectangle with:
  a) First corner: **0,0**
  b) Other corner: **420,297**, i.e. A3 paper size
  c) zoom–all if required.

6 Make layer VP current.

7 Menu bar with **View–Floating Viewports–1 Viewport** and:
  prompt   On/Off.../<First point> and enter: **10,10 <R>**
  prompt   Other corner and enter: **200,180 <R>**
  and   a viewport (A) is created with the model 3DWFM displayed as fig. (a).

8 Menu bar with **View–Floating Viewports–2 Viewports** and:
  prompt   Horizontal/<Vertical> and enter: **V <R>**
  prompt   Fit/<First corner> and enter: **10,200 <R>**
  prompt   Second point and enter: **300,280 <R>**
  and   two additional viewports (B and C) are created, each displaying the model in 3D – fig. (b).

9 Menu bar again with **View–Floating Viewports–4 Viewports** and:
  prompt   Fit/<First point> and enter: **220,10 <R>**
  prompt   Second point and enter: **@180,180 <R>**
  and   four new viewports (D,E,F,G) are created with the model displayed in each – fig. (c).

10 Final menu bar selection with **View–Floating Viewports–3 Viewports** (Right option) and:
prompt   First point and enter: **320,200 <R>**
prompt   Second point and enter: **@150,90 <R>**
and   three viewports (H,I,J) are created each with the model displayed as fig. (d). These viewports extend 'outside' the drawing paper border.

11 *What has been achieved?*
   a) ten viewports have been created
   b) these viewports have different sizes
   c) the viewports have been created in paper space
   d) each viewport displays the original 3DWFM model.

12 At the command line enter **MS <R>** and:
   a) toggled to model space
   b) the last viewport (top right) is active?
   c) each viewport displays the UCS icon at BASE?
   d) any viewport can be made active as before, i.e. move to required viewport and left-click.

13 Investigate the paper/model space toggle:
   a) from the command line with PS and MS
   b) status bar with double left-clicks on MODEL/PAPER
   c) menu bar with View–Paper Space or View–Model Space (Floating)
   d) decide on which toggle method you prefer.

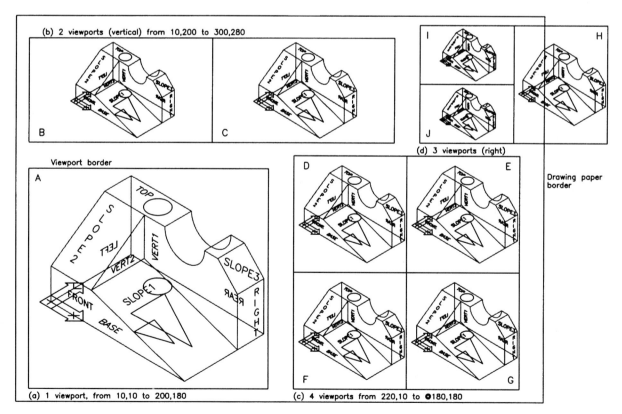

Figure 12.2   Creating the paper space viewports with 3DWFM.

Setting the viewpoints

1 Enter model space and set the following viewpoints in the named viewports:

viewport	viewpoint
A	SW Isometric
B	NE Isometric
C	NW Isometric
D	Right
E	Front
F	SE Isometric
G	Top
H	Bottom
I	Left
J	Back

2 The model is now displayed at a different viewpoint in each viewport. Centre each viewport about the point 100,60,50 at 250 magnification.

Adding text

1 In model space, make layer TEXT current and viewport (A) active.

2 Add an item of text using the following:
   a) start point: 0,–30
   b) height: 15 and rotation: 0
   c) text item: AutoCAD R14 (MS).

3 This item of text will be displayed in the ten viewports at different orientations due to the viewpoints. In some viewports (D,E,I,J) the text is viewed 'end-on'.

4 Enter paper space with PS <R>.

5 Add an item of text using:
   a) start point: centred on 210,145
   b) height: 20 and rotation: 0
   c) text: AutoCAD R14 <R>
      text: in <R>
      text: PAPER SPACE <R><R> – two returns.

6 At this stage your screen layout should resemble Fig. 12.3.

Modifying the layout

1 In paper space try and erase the model – you cannot.

2 In model space try and erase the paper space text – not possible.

3 In paper space, activate the ZOOM command and window viewport A. The viewport will nearly fill the screen and by entering model space, it is easier to work on the model.

4 In paper space, zoom previous.

5 In paper space, select the SCALE icon from the Modify toolbar and:

prompt	Select objects
respond	**completely window viewports H,I,J then right-click**
prompt	Base point
respond	**Intersection of lower left corner of viewport J**
prompt	Scale factor
enter	**0.75 <R>**.

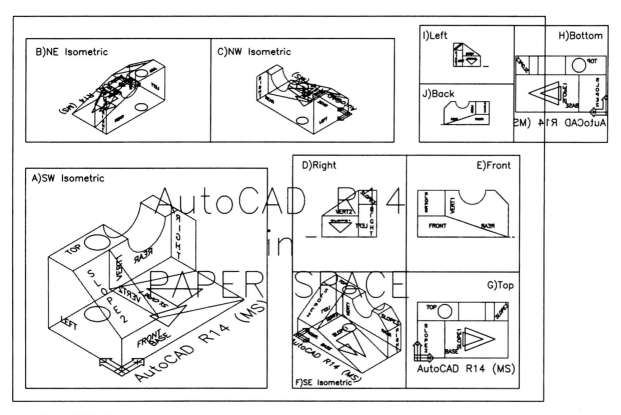

Figure 12.3 Working with the created paper space viewports.

6   With the MOVE icon:
    prompt   Select objects
    enter   **P <R><R>** – the previous option
    prompt   Base point and: **pick lower left corner as before**
    prompt   Second point and enter: @**–20,20 <R>**
    and   the three viewports are moved 'inside the paper'.

7   Enter model space and re-centre viewpoints H, I and J using the same centre point as before, i.e. 100,60,50 with 250 magnification.

8   Enter paper space and freeze the layer VP.

9   The 10 views of the model will be displayed with text but without the viewport borders – Fig. 12.4.

10 If you have access to a printer/plotter:
    a) print from any viewport in model space
    b) print from paper space.

11 This completes the exercise.

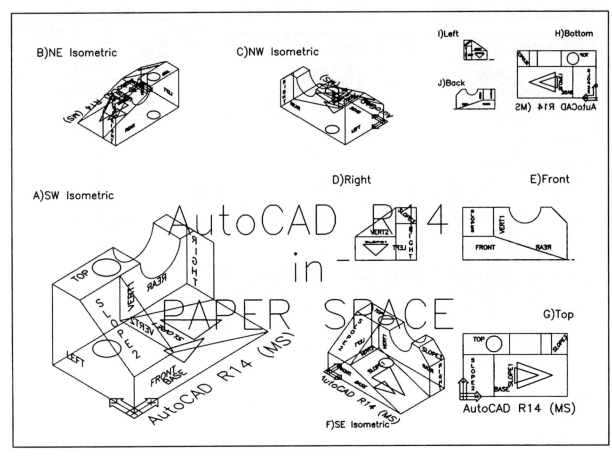

Figure 12.4 Completed paper space exercise with 3DWFM.

Comparison between model space and paper space

The following table gives a comparison between the model and paper space drawing environments:

Model space	*Paper space*
used to create the model	used to create the paper layout
model can be modified	model cannot be modified
tiled (fixed) viewports	untiled (floating) viewports
tilemode: 1	tilemode: 0
viewports restricted in size	viewports to any size
viewports 'fill screen'	viewports positioned to suit
viewports cannot be altered	viewports can be moved, copied, etc
cannot add viewports	additional viewports can be created
plot only active viewport	all viewports can be plotted
3D viewpoint	cannot use 3D viewpoint
WCS or UCS icon	paper space icon

A new 3D multiple viewport template file

Now that the model/paper space concept has been discussed, a new template file will be created which will allow all future models (surface and solid) to be displayed in multiple viewports. We will modify our existing 3DSTDA3 template file as it:
a) already has layers, e.g. MODEL, OBJECTS, TEXT, etc.
b) has a created dimension style – 3DSTD
c) other variables set.

Getting ready

1 Menu bar with **File–New** and:
   prompt   Create New Drawing dialogue box
   respond   1. pick **Use a Template**
                 2. scroll and pick **3DSTDA3.dwt**
                 3. pick **OK**.

2 The screen will display a black border at a SE Isometric viewpoint with the WCS at the left vertex of the border.

3 Erase the black border – we will not use it again.

4 Menu bar with **View–3D Viewpoint–Plan View–World UCS**.

5 Menu bar with **Format–Layer** and create two new layers:
name	*colour*	*linetype*
VP	number 22	continuous
SHEET	number 212	continuous

6 At the command line enter **TILEMODE <R>** and:
   prompt   New value for TILEMODE<1>
   enter   **0 <R>**.

7 Now entered the paper space environment – paper space icon.

Creating the drawing paper layout

1 Make layer SHEET current.

2 Menu bar with **Draw–Rectangle** and:
   prompt   First corner and enter: **0,0 <R>**
   prompt   Other corner and enter: **420,297 <R>** – A3 paper size.

3 Zoom–all and PAN to suit.

4 Draw a line from: 0,15 to: @420,0.

5 The area at the bottom of the 'paper' is for you to 'customize' as required.

Creating the viewports and setting the viewpoints

1 Make layer VP current.

2 Menu bar with **View–Floating Viewports–4 Viewports** and:
   prompt   Fit/<First point> and enter: **10,20 <R>**
   prompt   Second point and enter: **410,290 <R>**.

3 Screen displays four paper space viewports within the sheet border.

4   Enter model space with **MS <R>**.

5   Set the following viewpoints making the appropriate viewport active with the menu bar sequence **View–3D Viewpoint** and:

viewport   *viewpoint*
lower right   Top
upper right   Front
upper left   Right
lower left   SE Isometric

Setting three UCS positions

1   With the lower left viewport active, menu bar with **Tools–UCS–Origin** and:
prompt   Origin point and enter: **50,50,0 <R>**.

2   At the command line enter **UCSICON <R>** and:
prompt   On/OFF/...
enter   **A <R>** – for all viewports
then   **OR <R>** – icon at origin point in all viewports.

3   Menu bar with **Tools–UCS–Save** and:
prompt   ?/Desired UCS name
enter   **BASE <R>**.

4   Menu bar with **Tools–UCS–X Axis Rotate** and:
prompt   Rotation angle about X axis and enter: **90 <R>**.

5   Menu bar with **Tools–UCS–Save** and enter **FRONT** as the UCS name.

6   Menu bar with **Tools–UCS–Y Axis Rotate** and:
prompt   Rotation angle about Y axis and enter: **90 <R>**.

7   Save this UCS position as **RIGHT**.

8   These three UCS positions will assist with future model creation.

Finally

1   Restore UCS BASE, make layer MODEL current and make the lower left viewport active.

2   Menu bar with **File–Save As** and:
prompt   Save Drawing As dialogue box
respond   1. scroll at **Type**
             2. pick **Drawing Template File (*.dwt)**
prompt   list of existing template files
respond   1. enter File name as: **MV3DSTD**
             2. pick **Save**
prompt   Template Description dialogue box
respond   1. enter: **My multi-view 3D prototype layout drawing created on XX/YY/ZZ**
             2. pick **OK**.

3   This template file will be used continuously in all future exercises.

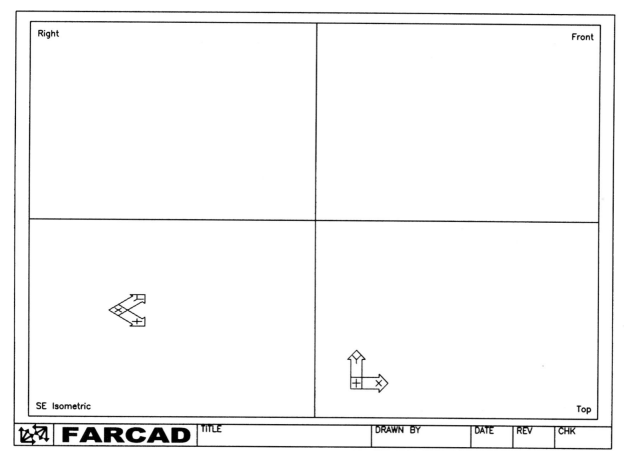

Figure 12.5 The MV3DSTD prototype A3 drawing sheet layout.

Checking the new MV3DSTD layout

Now that the MV3DSTD template file has been created we will add some 3D objects to 'check' the layout.

Note: in the exercise, try and reason out the coordinate entries.

1 Menu bar with **File–New** and:
prompt   Create New Drawing dialogue box
respond   1. pick **Use a Template**
            2. scroll and lick **MV3DSTD.dwt**
            3. pick **OK**.

2 The created multiple viewport drawing should be displayed with layer MODEL current, UCS BASE and lower left viewport active.

3 Display the surfaces toolbar.

4 Menu bar with **Draw–Surfaces–3D Surfaces** and:
prompt   3D Objects dialogue box
respond   **pick Box3d then OK**
prompt   Corner of box and enter: **0,0,0 <R>**
prompt   Length and enter: **200 <R>**
prompt   Cube/<Width> and enter: **100 <R>**
prompt   Height and enter: **80 <R>**
prompt   Rotation angle about Z axis and enter: **0 <R>**.

5 Select the WEDGE icon from the Surfaces toolbar and:
   prompt   Corner of wedge and enter: **0,0,0 <R>**
   prompt   Length and enter: **100 <R>**
   prompt   Width and enter: **100 <R>**
   prompt   Height and enter: **100 <R>**
   prompt   Rotation angle about Z axis and enter: **–90 <R>**.

6 Using the Properties icon from the Object Properties toolbar, change the colour of the wedge to blue.

7 Select the CONE icon from the Surfaces toolbar and:
   prompt   Base center point and enter: **70,50,80 <R>**
   prompt   Diameter/<radius> of base and enter: **50 <R>**
   prompt   Diameter/<radius> of top and enter: **0 <R>**
   prompt   Height and enter: **100 <R>**
   prompt   Number of segments<16> and ~~right-click~~. <R>

8 Change the colour of the cone to green.

9 Select the DISH icon from the Surface toolbar and:
   prompt   Center of dish and enter: **150,50,0 <R>**
   prompt   Diameter/<radius> and enter: **50 <R>**
   prompt   Number of longitudinal segments<16> and ~~right-click~~ <R>
   prompt   Number of latitudinal segments<8> and ~~right-click~~. <R>

10 Change the colour of the dish to magenta.

11 In each viewport, zoom-centre about the point **100,0,80** at **300** magnification – why these coordinates?

12 With UCS BASE make layer TEXT current and menu bar with **Draw–Text–Single Line Text** and:
   prompt   Start point and enter: **130,80,80 <R>**
   prompt   Height and enter: **10 <R>**
   prompt   Rotation and enter: **0 <R>**
   prompt   Text and enter: **AutoCAD <R><R>**.

13 Add two other text items using the following information:

	1	2
Named UCS	FRONT	RIGHT
start point	110,40,0	15,15,200
height	15	20
rotation	0	0
item	Release	14

14 Restore UCS BASE.

15 Enter paper space with **PS <R>**.

16 At the command line enter **DTEXT <R>** and enter the following:
   a) start point: centred on 210,155
   b) height: 10 and rotation: 0
   c) text: AutoCAD <R>
       text: Release <R>
       text: 14 <R>
       text: Paper <R>
       text: Space <R><R>.

17 Return to model space with **MS <R>**.

18 Menu bar with **View–Hide** in each viewport and your screen layout should resemble Fig. 12.6.

19 Try SHADE in each viewport – no model space text displayed?

20 This (long) chapter is now complete and we can concentrate on surface and solid modelling.

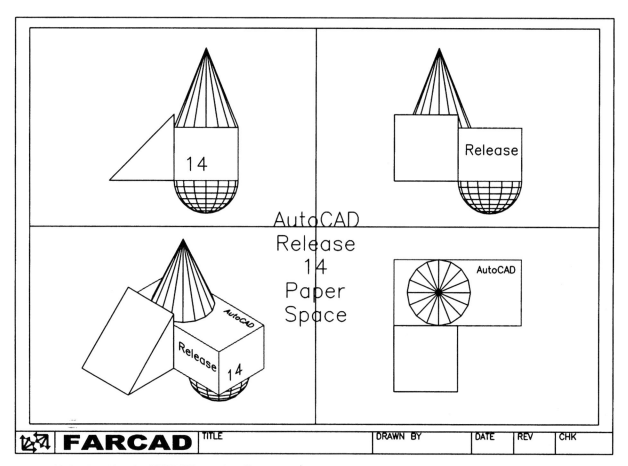

Figure 12.6 Checking the MV3DSTD template file.

Chapter 13

Surface modelling

The best way of describing a surface model is to think of a wire-frame model with 'skins' covering all the wires from which the model is constructed. The 'skins' convert a wire-frame model into a surface model with several advantages:

a) the model can be displayed with hidden line removal
b) there is no ambiguity
c) the model can be shaded and rendered.

AutoCAD Release 14 adds **FACETED** surfaces using a polygon mesh technique, but this mesh only approximates to curved surfaces. The mesh density (the number of facets) is controlled by certain system variables which will be discussed in the chapters which follow.

The different types of surface models available with Release 14 are:

- 3D faces
- 3D meshes
- polyface meshes
- ruled surfaces
- tabulated surfaces
- revolved surfaces
- edge surfaces
- 3D objects
- elevation/thickness surfaces.

Each surface type will be considered in a chapter on its own, with the exception of the elevation/thickness surfaces. This has already been discussed in Chapter 2 with 2.5D extrusions – did you realize that you were creating surface models at this early stage?

The surface commands can be activated:

a) from the menu bar with **Draw–Surfaces**
b) in icon form from the Surfaces toolbar
c) by direct keyboard entry, e.g. **3DFACE <R>**.

The various exercises will use all methods mentioned.

Chapter 14

3DFACE and PFACE

These two commands appear similar in operation, both adding faces (skins) to wire-frame models. If these added faces are in colour, the final model display can be quite impressive.

3Dface example

1 Menu bar with **File–New** and 'open' your **MV3DSTD** template file layer MODEL and UCS BASE. Display toolbars to suit.

2 Before creating the model, zoom–centre about 60,35,50 at 200 magnification in all viewports.

3 With the lower left (3D) viewport active, create a wire-frame model using the LINE command and the reference sizes in Fig. 14.1(a). The start point 1 should be at (0,0,0).

4 The created model has five 'surfaces' so make five new layers: F1 red, F2 blue, F3 green, F4 magenta and F5 yellow.

5 Make layer F1 current.

6 Menu bar with **Draw–Surfaces–3Dface** and:
prompt	First point and **pick Intersection icon of pt1**
prompt	Second point and **pick Intersection icon of pt2**
prompt	Third point and **pick Intersection icon of pt3**
prompt	Fourth point and **pick Intersection icon of pt4**
prompt	Third point and **right-click**.

7 Make layer F2 current and select the 3DFACE icon from the Surfaces toolbar and:
prompt	First point and pick Intersection pt2
prompt	Second point and pick Intersection pt3
prompt	Third point and pick Intersection pt5
prompt	Fourth point and pick Intersection pt2 then right-click.

8 Make layer F3 current and at the command line enter **3DFACE <R>** and:
prompt	First point and pick Intersection pt3
prompt	Second point and pick Intersection pt4
prompt	Third point and pick Intersection pt6
prompt	Fourth point and pick Intersection pt5 then right-click.

9 Use the 3DFACE command and add faces to:
a) face: 1256 with layer F4 current
b) face: 146 with layer F5 current.

10 Menu bar with **View–Hide** in each viewport – Fig. 14.1.

11 In each viewport **View–Shade–16 Color Filled** and:

viewport	colours displayed
top left	red, blue, green
top right	blue, red
lower right	blue
lower left	red, blue

12 Regen then save the drawing as **R14MOD\CHEESE**.

13 *Task*: *a*) alter the viewpoint to display the magenta and yellow surfaces of the model.
           b) can you reason out the zoom–centre coordinate values?

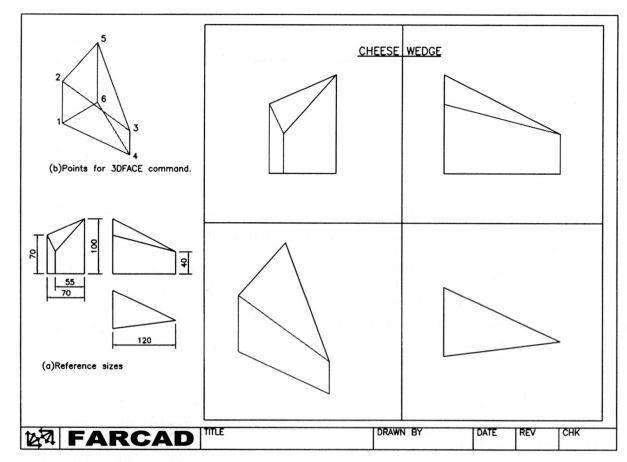

Figure 14.1 3DFACE example.

Explanation of the 3DFACE command

The 3DFACE command can be used to face any three or four sided 'surface'. The command allows 'continuous' faces to be created and will be demonstrated with a 2D example so:

1 With the screen display from the first example:
   a) erase the model – saved?
   b) make the lower right viewport active
   c) snap and grid on – set to 5 or 10
   d) refer to Fig. 14.2.

2 Activate the 3DFACE command and:

prompt	`First point` and pick a pt1
prompt	`Second point` and pick a pt2
prompt	`Third point` and pick a pt3 – 1st pt of next face
prompt	`Fourth point` and pick a pt4 – 2nd pt of next face
and	Face 1–2–3–4 displayed
prompt	`Third point` and pick pt5 – 3rd pt of face, 1st of next
prompt	`Fourth point` and pick a pt6 – 4th pt of face, 2nd of next
and	Face 3–4–5–6 displayed
prompt	`Third point` and pick a pt7 – 3rd pt of face, 1st of next
prompt	`Fourth point` and pick a pt8 – 4th of face, 2nd of next
and	Face 5–6–7–8 displayed
prompt	`Third point`
respond	pick points 9 and 10; 11 and 12; 13 and 14 in response to the third and fourth prompts then right-click
and	Faces 7–8–9–10; 9–10–11–12; 11–12–13–14 will be displayed as fig. (a).

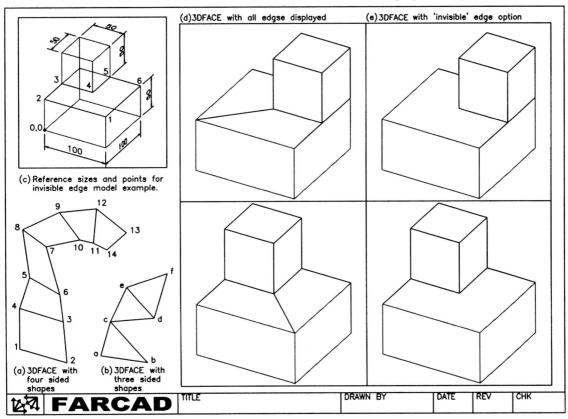

Figure 14.2 3DFACE example 2 – the INVISIBLE edge option.

3 Activate the 3DFACE command again and:

    prompt   First point and pick a pta

    prompt   Second point and pick a ptb

    prompt   Third point and pick a ptc – 1st of next face

    prompt   Fourth point and pick the same ptc – 2nd of next face

    and   Face a–b–c displayed

    prompt   Third point and pick a ptd – 3rd pt of face, 1st of next

    prompt   Fourth point and pick a pte – 4th pt of face, 2nd of next

    and   Face c–d–e displayed

    prompt   Third point and pick a ptf

    prompt   Fourth point and pick the same ptf then right-click

    and   Face d–e–f displayed as fig. (b).

4 Now erase the two complete 3D faces.

The invisible edge 3D FACE example

When the 3DFACE command is used with continuous four-sided 'shapes', all four sides of the face are displayed. It is possible to create a 3DFACE with an 'invisible edge'.

1 Refer to Fig. 14.2(c) and use the existing blank (?) four viewport configuration to:

    a) create a wire-frame model similar to that shown – easy?

    b) change the viewpoint in the viewports as follows:

        lower left: VPOINT 'R' at 300° and 30°

        upper left: VPOINT 'R' at 30° and 30°

        upper right and lower right: your own VPOINT 'R' values.

2 Set the object snap to INTERSECTION.

3 With layer F1 current, 3DFACE the four vertical sides of the lower part of the model.

4 With layer F3 current, 3DFACE the four vertical sides of the upper part of the model.

5 With layer F4 current, 3DFACE the top of the model.

6 With layer F2 current:

    a) activate the 3DFACE command and:

        prompt   First point and pick pt1

        prompt   Second point and pick pt2

        prompt   Third point and pick pt3

        prompt   Fourth point and pick pt4 then right-click.

    b) activate the 3DFACE command and:

        prompt   First point and pick pt1

        prompt   Second point and pick pt6

        prompt   Third point and pick pt5

        prompt   Fourth point and pick pt4 then right-click.

7 Menu bar with View–Hide and the model will be displayed with hidden line removal, with the edge between points 1 and 4 displayed – fig. (d).

8 Menu bar with View–Shade–16 Color Filled – impressive?

9 Menu bar with View–RegenAll and erase the two blue 3D faces by picking the lines joining points 1 and 4.

10 Still with layer F2 current:
    a) activate the 3DFACE command and:
      prompt   First point and pick pt1
      prompt   Second point and pick pt2
      prompt   Third point and pick pt3
      prompt   Fourth point
      enter   **I <R>** – the invisible edge option
      then   pick pt4
      prompt   Third point and right-click
      and   the 3D face is displayed without edge 1–4.
    b) 3DFACE again and:
      prompt   First point and pick pt1
      prompt   Second point and pick pt6
      prompt   Third point and pick pt5
      prompt   Fourth point
      enter   **I <R>**
      and   pick pt4
      prompt   Third point and right-click
      and   the 3D face is displayed without edge 1–4.

11 Hide and shade – fig. (e).

12 Cancel the object snap intersection mode.

13 Save if required, but it will not be used again.

PFACE

A PFACE is a polygon mesh and is similar to a 3DFACE. It allows the user to define a number of vertices for the surface to be faced, not the 3 or 4 allowed with the 3DFACE command. The following example has a rather long set of prompts:

1 Open your MV3DSTD template file and:
    a) make four new layers: F1 blue, F2 green, F3 magenta, F4 cyan
    b) layer MODEL current and restore UCS FRONT
    c) lower left viewport (3D) active.

2 Set the elevation to 0 and the thickness to −200.

3 Select the POLYGON icon from the Draw toolbar and:
    a) number of sides: 5
    b) centre of polygon: 0,0
    c) circumscribed circle radius: 50.

4 Zoom–centre about the point 0,0,−100 at 225 magnification.

5 Refer to Fig. 14.3 which only considers the 3D viewport.

6 Menu bar with **View–Hide** to display the pentagonal prism without a 'top' – fig. (a).

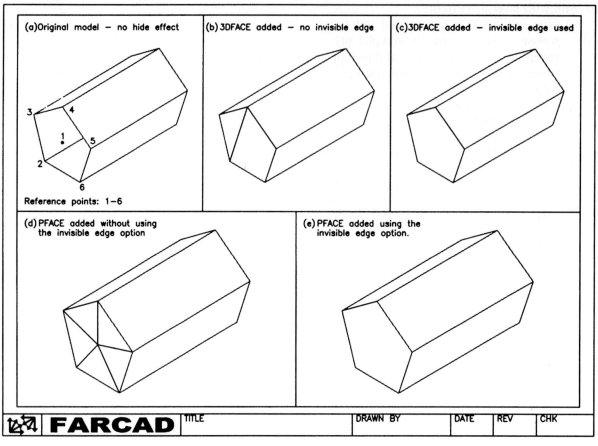

Figure 14.3 3DFACE example.

7 With layer F1 current:

a) select the 3DFACE icon and:

First point	Intersection pt2
Second point	Intersection pt3
Third point	Intersection pt4
Fourth point	Intersection pt4
Third point	right-click

b) select the 3DFACE icon and:

First point	Intersection pt2
Second point	Intersection pt6
Third point	Intersection pt5
Fourth point	Intersection pt4
Third point	right-click.

8 Hide the model to display the 'top surface' as fig. (b).

9 Make layer F2 current and freeze layer F1 then:

a) activate the 3DFACE command and:

First point	Intersection pt2
Second point	Intersection pt3
Third point	Intersection pt4
Fourth point	enter **I <R>** then intersection pt4
Third point	right-click

b) 3DFACE again and:

First point	Intersection pt2
Second point	Intersection pt6
Third point	Intersection pt5
Fourth point	enter **I <R>** then intersection pt4
Third point	right-click.

10 Hide to display the model with the 3D face invisible edge option as fig. (c).

11 Make layer F3 current and freeze layer F2.

12 At the command line enter **PFACE <R>** and:

prompt	Vertex 1 and enter: **0,0 <R>**, i.e. pt1
prompt	Vertex 2 and pick Intersection pt2
prompt	Vertex 3 and pick Intersection pt3
prompt	Vertex 4 and pick Intersection pt4
prompt	Vertex 5 and pick Intersection pt5
prompt	Vertex 6 and pick Intersection pt6
prompt	Vertex 7 and right-click: no more vertices
prompt	Face 1,vertex 1 and enter: **1 <R>**, i.e. pt1
prompt	Face 1,vertex 2 and enter: **2 <R>**
prompt	Face 1,vertex 3 and enter: **3 <R>**
prompt	Face 1,vertex 4 and right-click, i.e. end of face 1
prompt	Face 2,vertex 1 and enter: **1**
prompt	Face 2,vertex 2 and enter: **3**
prompt	Face 2,vertex 3 and enter: **4**
prompt	Face 2,vertex 4 and right-click, i.e. end of face 2
prompt	Face 3,vertex 1 and enter: **1**
prompt	Face 3,vertex 2 and enter: **4**
prompt	Face 3,vertex 3 and enter: **5**
prompt	Face 3,vertex 4 and right-click, i.e. end of face 3
prompt	Face 4,vertex 1 and enter: **1**
prompt	Face 4,vertex 2 and enter: **5**
prompt	Face 4,vertex 3 and enter: **6**
prompt	Face 4,vertex 4 and right-click, i.e. end of face 4
prompt	Face 5,vertex 1 and enter: **1**
prompt	Face 5,vertex 2 and enter: **6**
prompt	Face 5,vertex 3 and enter: **2**
prompt	Face 5,vertex 4 and right-click, i.e. end of face 5
prompt	Face 6,vertex 1 and right-click to end command.

13 Menu bar with View–Hide to display the end of the prism with a pface surface – fig. (d).

14 Make layer F4 current and freeze layer F3.

15 Repeat the PFACE command line entry and:

prompt	Vertex 1 and enter: **0,0 <R>**
prompt	Vertex 2 and pick Intersection pt2
prompt	Vertex 3 and pick Intersection pt3
prompt	Vertex 4 and pick Intersection pt4
prompt	Vertex 5 and pick Intersection pt5
prompt	Vertex 6 and pick Intersection pt6
prompt	Vertex 7 and right-click
prompt	Face 1,vertex 1 and enter: **–1 <R>**
prompt	Face 1,vertex 2 and enter: **2 <R>**
prompt	Face 1,vertex 3 and enter; **–3 <R>**

prompt	Face 1,vertex	4 and right-click
prompt	Face 2,vertex	1 and enter: −**1**
prompt	Face 2,vertex	2 and enter: **3**
prompt	Face 2,vertex	3 and enter: −**4**
prompt	Face 2,vertex	4 and right-click
prompt	Face 3,vertex	1 and enter: −**1**
prompt	Face 3,vertex	2 and enter: **4**
prompt	Face 3,vertex	3 and enter: −**5**
prompt	Face 3,vertex	4 and right-click
prompt	Face 4,vertex	1 and enter: −**1**
prompt	Face 4,vertex	2 and enter: **5**
prompt	Face 4,vertex	3 and enter: −**6**
prompt	Face 4,vertex	4 and right-click
prompt	Face 5,vertex	1 and enter: −**1**
prompt	Face 5,vertex	2 and enter: **6**
prompt	Face 5,vertex	3 and enter: −**2**
prompt	Face 5,vertex	4 and right-click
prompt	Face 6,vertex	1 and right-click.

16 Now View–Hide to display the model with the hidden edge option of the pface command.

17 Save if required, but this model will not be used again.

Summary

1 The 3DFACE and PFACE commands allow **surface models** to be created by drawing 'skins' over wire-frame models.
2 The HIDE command allows surface models to be displayed with hidden line removal.
3 The 3DFACE command can only be used with three- or four-sided 'shapes' although continuous faces can be created.
4 The PFACE command can be used with multi-sided figures.
5 Both commands have an invisible edge option.
6 The commands can be activated:
   a) 3DFACE: command line, menu bar or icon
   b) PFACE: command line only.
7 Using separate coloured layers for different faces allows models to be displayed in colour using the SHADE command.

Assignment

A surface model is to be created in Activity 10, so:

1 Use MV3DSTD template file, layer MODEL and UCS BASE current.

2 Create a wire-frame model of the hexagonal prism using the sizes given and zoom-centre about 0,0,75 at 200 magnification.

3 Make two new coloured layers:
verticals: blue
slopes: green

4 Use the 3DFACE command with the coloured layers to convert the wire-frame model into a surface model.

5 The base has not been surfaced – up to you?

6 Freeze layer MODEL, then hide and shade.

7 Save?

Chapter 15

3DMESH

A 3D mesh (or more correctly, a 3D polygon mesh) consists of a series of 3D faces in a rectangular array pattern. This mesh matrix is defined by **M × N vertices** and:
a) M is the number of 'columns' in the *x* direction
b) N is the number of 'rows' in the *y* direction
c) the user enters the *x, y, z* coordinates of every vertex in the matrix.

3D mesh example

1 Open your MV3DSTD template file with layer MODEL, UCS BASE and the lower right viewport active.

2 Display the Surfaces toolbar and refer to Fig. 15.1.

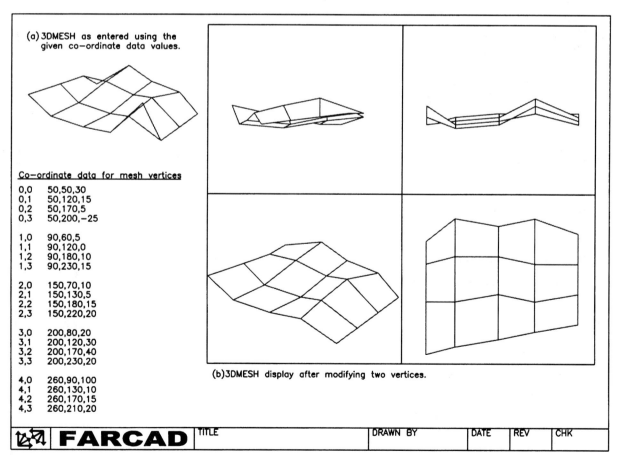

(a) 3DMESH as entered using the given co-ordinate data values.

Co-ordinate data for mesh vertices

0,0	50,50,30
0,1	50,120,15
0,2	50,170,5
0,3	50,200,−25
1,0	90,60,5
1,1	90,120,0
1,2	90,180,10
1,3	90,230,15
2,0	150,70,10
2,1	150,130,5
2,2	150,180,15
2,3	150,220,20
3,0	200,80,20
3,1	200,120,30
3,2	200,170,40
3,3	200,230,20
4,0	260,90,100
4,1	260,130,10
4,2	260,170,15
4,3	260,210,20

(b) 3DMESH display after modifying two vertices.

FARCAD | TITLE | | DRAWN BY | DATE | REV | CHK

Figure 15.1 3DMESH example.

3 Zoom–centre about the point 150,150,25 at 250 magnification.

4 Select the 3DMESH icon from the Surfaces toolbar and:
   prompt   Mesh M size and enter: **5 <R>**
   prompt   Mesh N size and enter: **4 <R>**
   prompt   Vertex (0,0) and enter: **50,50,30 <R>**
   prompt   Vertex (0,1) and enter: **50,120,15 <R>**
   prompt   Vertex (0,2) and:
   respond   refer to Fig. 15.1 and enter the required coordinates in response to the vertex prompts.

5 When the last vertex coordinate is entered – Vertex (4,3), the 3D mesh will be displayed as fig. (a).

6 Two of the vertices have been entered wrongly, these being:
   vertex (0,3): 50,200,–25
   vertex (4,0): 260,90,100

7 Menu bar with **Modify–Object–Polyline** and:
   prompt   Select polyline
   respond   **pick any point on the mesh**
   prompt   Edit vertex/Smooth surface/...
   enter   **E <R>** – the edit vertex option
   prompt   Vertex (0,0). Next/Previous/...
   and   an X is displayed at vertex (0,0) in all viewports
   enter   **N <R>** until X at vertex (0,3) and:
   prompt   Vertex (0,3). Next/Previous...
   enter   **M <R>** – the move option
   prompt   Enter new location
   enter   **50,200,5 <R>** – absolute entry
   prompt   Vertex (0,3). Next/Previous...
   enter   **N <R>** until X at vertex (4,0) and:
   prompt   Vertex (4,0). Next/Previous/...
   enter   **M <R>**
   prompt   Enter new location
   enter   **@0,0,–95 <R>** – relative entry
   prompt   Vertex (4,0). Next/Previous/...
   enter   **X <R>** – to exit edit vertex option
   prompt   Edit vertex/Smooth surface/...
   enter   **X <R>** – to exit command.

8. The mesh will be displayed as fig. (b), with the two wrong vertices have been 'repositioned'.

9 Use the hide command.

10 Save if required, as this exercise is now complete.

Notes

1 This example has been a brief introduction to the 3DMESH command.

2 The command requires the user to enter all vertex values as coordinates and is therefore very tedious to use. You can also reference existing objects if these are displayed.

3 The EDIT POLYLINE command can be used with 3D meshes. This is the same as the 2D poly edit command.

4 Meshes can be drawn in 3D with x, y, z coordinates or in 2D with x and y coordinates.

5 The mesh values M and N can be between 2 and 256.

6 A 3D mesh is a single object.

7 The command can be activated:
   a) by icon selection from the Surfaces toolbar
   b) from the menu bar with Draw–Surfaces
   c) by entering 3DMESH at the command line.

Chapter 16

Ruled surface

A ruled surface is a polygon mesh created between two defined boundaries selected by the user. The objects which can be used to define the boundaries are lines, arcs, circles, points and 2D/3D polylines. The surface created is a 'one-way' mesh of straight lines drawn between the two selected boundaries.

The ruled surface effect will be demonstrated by worked examples, the first being to allow the user to become familiar with the basic terminology.

Example 1

1 Begin a new 2D drawing and create two layers, MOD red and RULSUR blue. Refer to Fig. 16.1.

2 Display the Draw, Modify and Surfaces toolbars.

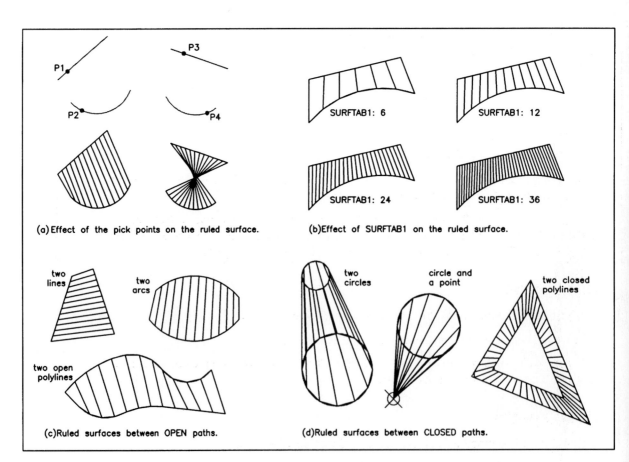

(a) Effect of the pick points on the ruled surface.

(b) Effect of SURFTAB1 on the ruled surface.

(c) Ruled surfaces between OPEN paths.

(d) Ruled surfaces between CLOSED paths.

Figure 16.1 Ruled surface example 1 – usage and basic terminology.

Pick points effect

1 With layer MOD current, draw two lines and two arcs as fig. (a).

2 Make layer RULSUR current.

3 Select the RULED SURFACE icon from the Surfaces toolbar and:

prompt	Select first defining curve
respond	**pick a point P1 on the first line**
prompt	Select second defining curve
respond	**pick a point P2 on the first arc**
and	a blue ruled surface is drawn between the two objects.

4 Menu bar with **Draw–Surfaces–Ruled Surface** and:
   a) first defining curve prompt: pick point P3 on second line
   b) second defining curve prompt: pick point P4 on second arc
   c) ruled surface drawn between the line and arc.

5 The ruled surface drawn between selected objects is dependent on the pick point positions.

Effect of the SURFTAB1 system variable

1 With layer MOD current draw a line and arc as fig. (b).

2 Copy the line and arc to three other places on screen.

3 Make layer RULSUR current.

4 At the command line enter **SURFTAB1 <R>** and:
prompt	New value for SURFTAB1<?>
enter	**6 <R>**.

5 At the command line enter **RULESURF <R>** and:
prompt	Select first defining curve
respond	**pick a point on the first line**
prompt	Select second defining curve
respond	**pick a point on the first arc**.

6 By entering SURFTAB1 at the command line, enter new values of 12, 24 and 36 and add a ruled surface between the other lines and arcs.

7 The system variable **SURFTAB1** controls the display of the ruled surface effect, i.e. the number of 'strips' added between the selected boundaries. The default value is 6. The value of SURFTAB1 is dependent on the 'size' of the boundary objects.

Open paths

1 An open path is defined as a line, arc or open polyline.

2 With layer MOD current draw some open paths as fig. (c).

3 Using the ruled surface command and with SURTFAB1 set to your own value, add ruled surfaces between the drawn open paths.

Closed paths

1 A closed path is defined as a circle, point or closed polyline.

2 Draw some closed paths as fig(d) and add ruled surfaces between them.

Note

1 A ruled surface can only be drawn:
   a) between TWO OPEN paths
   b) between TWO CLOSED paths.

2 A ruled surface **cannot** be created between an open and a closed path. If a line and a circle are selected as the defining curves, the following message will be displayed:

   Cannot mix closed and open paths.

3 This exercise is now complete and need not be saved.

Example 2

1 Open your MV3DSTD template file and refer to Fig. 16.2. Note that I have only displayed the 3D viewport.

2 With layer MODEL current and UCS BASE, create the model base from lines and arcs (trimmed circles?) using the sizes given in fig. (a).

3 Make a new layer RULSRF, colour blue and current.

4 Set the SURFTAB1 system variable to 18.

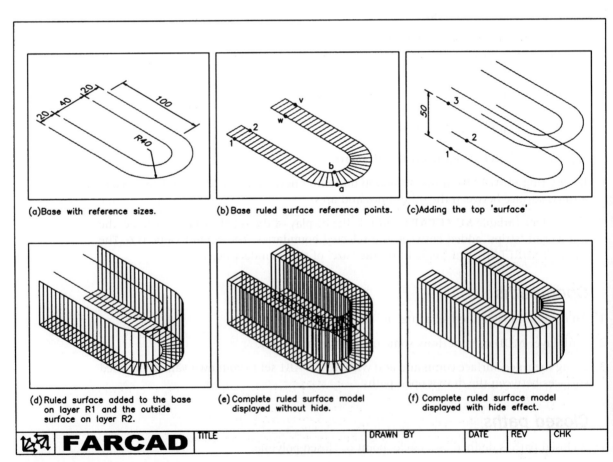

(a) Base with reference sizes.

(b) Base ruled surface reference points.

(c) Adding the top 'surface'

(d) Ruled surface added to the base on layer R1 and the outside surface on layer R2.

(e) Complete ruled surface model displayed without hide.

(f) Complete ruled surface model displayed with hide effect.

FARCAD

TITLE		DRAWN BY	DATE	REV	CHK

Figure 16.2 Ruled surface example 2.

5 Using the ruled surface icon from the Surfaces toolbar, select the following defining curves:
   a) lines 1 and 2
   b) arcs a and b
   c) lines v and w
   d) effect as fig. (b).

6 Erase the ruled surfaces and create the top surface of the model by copying the base objects:
   a) from the point 0,0,0
   b) by @0,0,50 – fig. (c).

7 With layer RULSRF still current, select the ruled surface icon and select the following defining curves as fig. (c).
   a) lines 1 and 2 – ruled surface added
   b) lines 3 and 1 – no ruled surface added and following message displayed:

   Object not usable to define ruled surface – why?

   c) Explanation: when the second set of defined curves were being selected:
      i) point 3 was picked satisfactorily
      ii) point 1 could not be picked – you were picking the previous ruled surface added between lines 1 and 2.
   d) cancel the ruled surface command (ESC) and erase the added ruled surface.

8 Make the following four new layers:
   R1 – red; R2 – blue; R3 – green; R4 – magenta.

9 *a*) Make layer R1 current
   b) Add a ruled surface to the base of the model (three needed).

10 *a*) Make layer R2 current
   b) Freeze layer R1
   c) Add a ruled surface to the three 'outside' vertical surfaces of the model
   d) Thaw layer R1 – fig. (d).

11 *a*) Make layer R3 current
   b) Freeze layers R1 and R2
   c) Rule surface the top surface of the model.

12 *a*) Make layer R4 current and freeze layer R3
   b) Add a ruled surface to the three 'inside' vertical surfaces.

13 *a*) Thaw layers R1, R2 and R3
   b) Model displayed a fig. (e).

14 Menu bar with **View–Hide** to give fig. (d). (f)

15 Menu bar with **View–Shade–16 Color Filled** – impressive?

16 RegenAll and save the drawing as **R14MOD\FLOWBED** – it may be used in a later exercise.

17 *Note.*
   When the ruled surface command is being used with adjacent surfaces, it is recommended that:
   a) a layer be made for each ruled surface to be added
   b) once a ruled surface has been added, the layer should be frozen before the next surface is added
   c) the new surface layers should be coloured for effect.

18 *Task.*

   a) Display the model at various viewpoints to 'see' the base

   b) The original model was created from lines and circles/arcs. The base could have been created from a polyline and then offset and copied. Try this and add a ruled surface and note that only one set of defining curves is required. What about SURFTAB1?

Example 3

This example will investigate how a ruled surface can be added to a surface which has a circular/slotted hole in it. The example will be in 2D, but the procedure is identical for a 3D model.

1 Begin a new 2D drawing and refer to Fig. 16.3.

2 Make two new layers, MOD red (current) and RULSRF blue.

3 Set SURFTAB1 to 24.

4 Using the LINE icon draw a square of side 60 with a 15 radius circle at the square 'centre' – snap on helps.

5 Using the Ruled Surface icon, pick any line of the square and the circle as the defining curves. No ruled surface can be added because of the open/closed path effect – fig. (a).

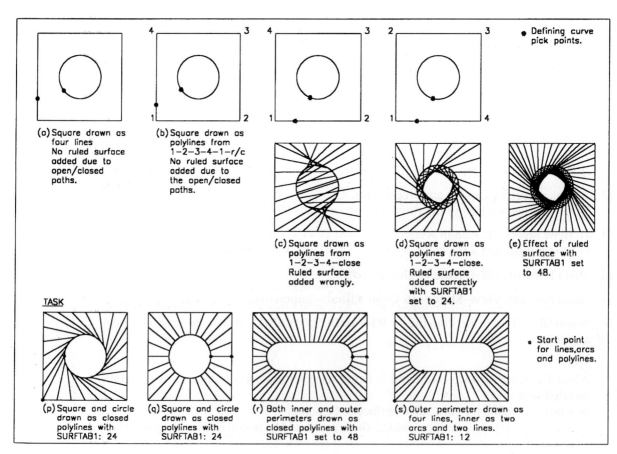

Figure 16.3 Ruled surface example 3.

6 With the Polyline icon, draw a 60 sided square from 1–2–3–4–1–<R> as fig. (b) and draw the 15 radius circle. Add a ruled surface and the open/closed path message is displayed and no ruled surface is added.

7 Draw a **closed** polyline square using the points 1–2–3–4 in the order given in fig. (c). Draw the circle. With the ruled surface icon pick the defining curves indicated and the a ruled surface is added, but not as expected.

8 Draw a 60 sided square as a **closed** polyline using points 1–2–3–4 in the order given in fig. (d). Draw the circle then add a ruled surface picking the defining curves indicated. The added ruled surface is not quite 'correct' at the circle (hole).

9 Erase the ruled surface effect, set SURFTAB1 to 48 and repeat the ruled surface command – fig. (e). Set SURFTAB1 back to 24.

10 *Note*.
   a) When a ruled surface is added between two defined curves, the surface 'begins at the defined curve start points'. It is thus essential that the defined curves:
      i) are **drawn in the same direction**
      ii) are **drawn from the same 'relative' start point**.
   b) Circular holes require to be drawn as two closed polyarcs.

11 *Task*.
   Using the information given in step 10, add ruled surfaces to the following models:
   a) Figure (p): square drawn as a closed polyline and circle drawn as two closed polyarcs. Note start points.
   b) Figure (q): square drawn as a closed polyline and circle drawn as two closed polyarcs. Note that the start points differ from those in fig. (p).
   c) Figure (r): both the outer and inner perimeters are drawn as closed polylines/polyarcs. Note the start points.
   d) Figure (s): the outer perimeter is drawn as four lines, and the inner as two arcs and two lines.

12 When this task is complete, this exercise is finished. Save?

Example 4

A ruled surface is one of the most effective surface modelling commands, and I have included another 3D model to demonstrate how it is used. The procedure when adding a ruled surface is basically the same with all models, this being:
a) create the 3D wire-frame model
b) make new coloured layers for the surfaces to be added
c) use the ruled surface command with layers current as required.

1 Open your MV3DSTD template file and refer to Fig. 16.4.
2 Make four new layers, R1 red, R2 blue, R3 green and R4 magenta.
3 With layer MODEL current, restore UCS FRONT and make the lower left (3D) viewport active.

4 Select the POLYLINE icon and draw:
   From point   0,0
   To point   @0,100
   To point   Arc option, i.e. A <R>
   Arc endpt   @50,50 then right-click.

5 Centre each viewport about the point 50,75,0 at 175 magnification.

6 Offset the polyline by 20 'inwards'.

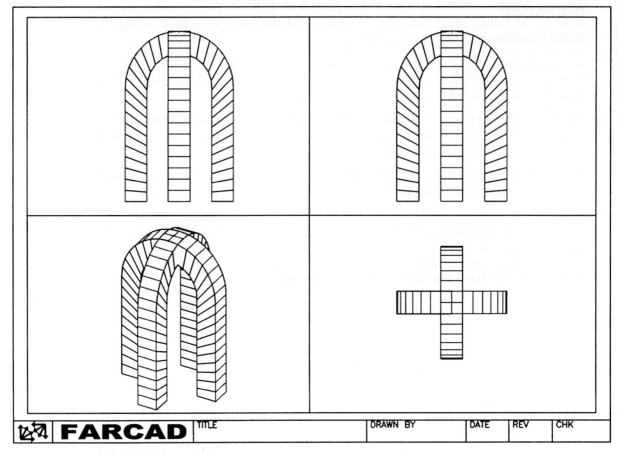

Figure 16.4 Ruled surface example 4 – ARCHES.

7 Copy the two polylines from: 0,0, by: @0,0,−20.

8 Change the viewpoint in the lower left viewport with the rotate option and angles:
first prompt   300
second prompt   30

9 Set SURFTAB1 to 18.

10 Making each layer R1–R4 current, add a ruled surface to each 'side; of the model, remembering to freeze layers as in the second example.

11 Restore UCS BASE and polar array the complete model (crossing selection) using:
a) centre point: 50,10
b) number of items: 4
c) angle: 360 with rotation.

12 Hide, shade, etc. – impressive result?

13 Save the complete model as **R14MOD\ARCHES** for future recall.

14 *Note*: the top 'square' of the arrayed arches – comments?

Summary

1 A ruled surface can be added between lines, circles, arcs, points and polylines.
2 The command can be activated in icon form, from the menu bar or by keyboard entry.
3 The command can be used in 2D or 3D.
4 A ruled surface **can only** be added between:
   a) two open paths, e.g. lines, arcs, polylines (not closed)
   b) two closed paths, e.g. circles, points, closed polylines.

5 With closed paths, the correct effect can only be obtained if:
   a) the paths are drawn in the same direction
   b) the paths start at the 'same relative point'.

6 The system variable SURFTAB1 controls the number of ruled surface 'strips' added between the two defining curves. The default is 6.

Assignment

This activity is interesting to complete, the procedure being the same as in the two examples.

Activity 11: Ornamental flower bed

1 Open your MV3DSTD template file.

2 Create the wire-frame model using the sizes given and note that the 'vertical arch' requires the UCS RIGHT to be current. Use your discretion for any sizes omitted.

3 Make four coloured layers.

4 Add ruled surfaces to the 'four sides' of the model using the four new layers correctly. Optimize the value of SURFTAB1.

5 Zoom centre about 90,50,50 at ?? magnification.

6 Hide, shade, save.

Chapter 17

Tabulated surface

A tabulated surface is a parallel polygon mesh created along a path, the user defining:
a) the **path curve** – the profile of the final model
b) the **direction vector** – the 'depth' of the profile.

The path curve can be created from lines, arcs, circles, ellipses, splines or 2D/3D polylines, The direction vector **must** be a line or an open 2D/3D polyline. The system variable SURFTAB1 determines the 'appearance' of the tabulated surface along curves.

Example

1 Open your MV3DSTD template file with layer MODEL current, lower left viewport active, toolbars to suit.

2 Refer to Fig. 17.1 and draw two lines:
  a) from: 0,0,0 to: @0,0,80
  b) from: 0,0,0 to: @−80,0
  Note: only the 3D viewport is displayed in Fig. 17.1.

3 With UCS BASE draw a closed polyline using:
  From: 50,50   To: @50,0   To: @0,20   To: @−20,0
  To: 0,30   To: @50,0   To: @0,20   To: @−80,0
  To: close.

4 Restore UCS RIGHT and draw another closed polyline using:
  From: 50,50   To: @100,0   To: @0,70   To: @−40,0
  To: @0,−50   To: @−60,0   To: close.

5 Fillet this polyline shape with a radius of 5.

6 Restore UCS BASE and zoom centre about 0,60,60 at 250 mag.

7 Select the TABULATED SURFACE icon from the Surfaces toolbar and:
  prompt   Select path curve
  respond   **pick polyline 1 as fig. (a)**
  prompt   Select direction vector
  respond   **pick line 1 as indicated**
  and   a tabulated surface is added to the path curve.

8 The added tabulated surface has a 'depth' equal to the length of the direction vector.

9 Set SURFTAB1 to 6 – probably is?

10 Menu bar with **Draw–Surfaces–Tabulated Surface** and:
  prompt   Select path curve and pick polyline 2
  prompt   Select direction vector and pick line 2 as indicated.

11 Figure 17.1 displays the results of the tabulated surface commands:
  a) reference information
  b) tabulated surfaces without hide
  c) tabulated surfaces with hide
  d) at a NW Isometric viewpoint.

12 *Task*.
  a) Erase the tabulated surfaces.
  b) Repeat the tabulated surface commands, but pick the direction vector lines at the 'opposite ends' from the exercise. The path curve will be 'extruded' in the opposite sense.

13 *Investigate*.
  a) Erase all objects – save?
  b) With UCS BASE draw two polylines:
     i) from: 0,0 to: @0,100 to: arc option to @–50,50 to: <R>
     ii) from: 100,0 to: @0,100 to: @50,50 to: @0,50 to: <R>.
  c) Restore UCS FRONT and draw two 15 radius circles with centres at 30,0 and 130,0
  d) Set SURFTAB1 to 18.
  e) Using the tabulated surface command twice:
     i) pick the circles as the path curve
     ii) pick the polylines as the direction vector.
  f) Comment on the result – think angles?

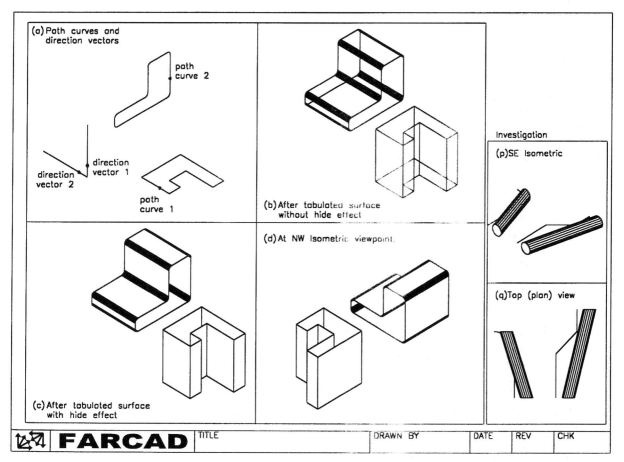

Figure 17.1 Tabulated surface example.

Summary

1 A tabulated surface is a parallel polygon mesh.

2 The command requires:
   a) a path curve – a single object
   b) a direction vector – generally a line.

3 The command can be used in 2D or 3D.

4 The final surface orientation is dependent on the direction vector 'pick point'.

5 SURFTAB1 determines the surface appearance with curved objects.

6 The command can be activated:
   a) in icon form from the Surfaces toolbar
   b) from the menu bar with Draw–Surfaces
   c) by entering **TABSURF <R>** at the command line.

Chapter 18

Revolved surface

A revolved surface is a polygon mesh generated by rotating a profile about an axis, the user selecting:
a) the **path curve** – a single object, e.g. a line, arc, circle or 2D/3D polyline
b) the **axis of revolution** – generally a line, but can be an open or closed polyline.

The generated mesh is controlled by two system variables:
a) SURFTAB1: controls the mesh in the direction of the revolution
b) SURFTAB2: defines any curved elements in the profile
The default value for both variables is 6.

Example 1

1 Open the MV3DSTD template file, layer MODEL current, UCS BASE and refer to Fig. 18.1.

2 Make the lower right viewport active and display toolbars.

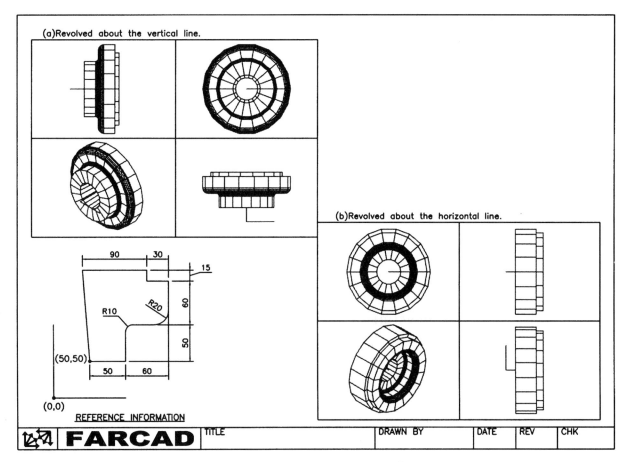

Figure 18.1 Revolved surface example 1.

3 Draw two lines:
   a) from: 0,0 to: @100,0
   b) from: 0,0 to: @0,100.

4 Set SURFTAB1 to 18 and SURFTAB2 to 6 – command line entry.

5 Using the polyline icon from the Draw toolbar, create a **CLOSED** polyline shape using the reference sizes given in Fig. 18.1. The start point is to be 50,50.

6 Select the REVOLVED SURFACE icon from the Surfaces toolbar and:
   prompt   Select path curve
   respond   **pick any point on the polyline**
   prompt   Select axis of revolution
   respond   **pick the vertical line**
   prompt   Start angle<0>
   enter   **0 <R>**
   prompt   Included angle (+ = ccw, - = cw)<Full circle>
   enter   **360 <R>**.

7 A revolved surface model will be displayed in each viewport.

8 Zoom centre about 0,120 at 350 magnification.

9 Hide each viewport – fig. (a).

10 Erase the revolved surface (regen needed?) and from the menu bar select **Draw–Surfaces–Revolved Surface** and
   a) path curve: pick the polyline shape
   b) axis of revolution: pick the horizontal line
   c) start angle: 0
   d) included angle: 360.

11 Zoom centre about 100,0 at 400 magnification.

12 Hide the viewports – fig. (b).

13 Save if required.

Example 2

1 Open the MV3DSTD template file, layer MODEL, UCS BASE with the the lower right viewport active and refer to Fig. 18.2.

2 Draw a line from 0,0 to @0,250.

3 With the polyline icon, draw an **OPEN** polyline shape using the sizes in fig. (a) as a reference. The start point is to be (0,50) but the final polyline shape is at your discretion – it is your wine glass design.

4 Set SURFTAB1 to 18 and SURFTAB2 to 6.

5 At the command line enter **REVSURF <R>** and:
   a) path curve: pick the polyline shape
   b) axis of revolution: pick the line
   c) start angle: enter 0
   d) included angle: enter 270.

6 Set the following 3D viewpoints in the named viewports:
   Top left: NE Isometric   Top right: NW Isometric
   Lower left: SE Isometric   Lower right: SW Isometric.

7 Zoom centre about 0,120 at 200 magnification.

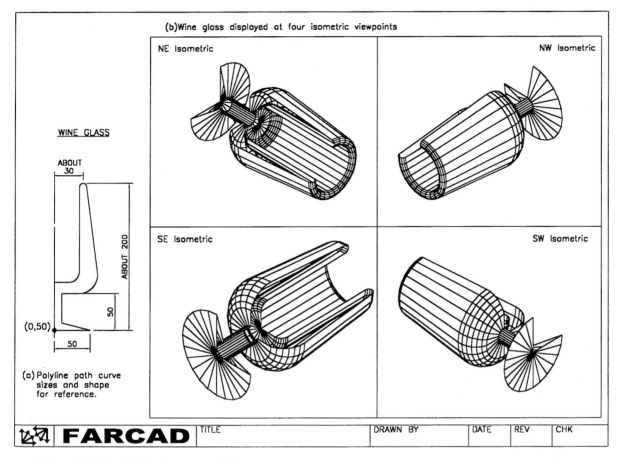

Figure 18.2 Revolved surface example 2.

8 Hide the viewports – fig. (b).

9 Save if required.

Summary

1 The revolved surface command can be used to produce very complex surface models from relatively simple profiles.

2 The resultant polygon mesh is controlled by the two system variables SURFTAB1 and SURFTAB2:
SURFTAB1: controls the mesh in the direction of rotation
SURFTAB2: controls the display of curved elements in the profile.

3 The start angle can vary between 0 and 360. A start angle of 0 means that the surface is to begin on the current drawing plane. This is generally what is required.

4 The included angle allows the user to define the angle the path curve is to be revolved through. The 360 default value gives a complete revolution, but 'cut-away' models can be obtained with angles less than 360°.

5 The direction of the revolved surface is controlled by the sign of the included angle and:
a) +ve for anti-clockwise revolved surfaces
b) −ve for a clockwise revolution.

6 The command can be activated by icon, from the menu bar or by command line entry.

Assignment

This activity requires two profiles to be created and revolved about two different axes. Adding colour to the revolved surfaces greatly enhances the model appearance with shading and rendering.

Activity 12: Garden furniture set

1 Use your MV3DSTD template file.

2 Make the lower right viewport active and restore UCS FRONT.

3 Draw two polyline profiles using the reference data given. Use your discretion for sizes not given, or design your own table and chair.

4 Set SURFTAB1 to 18 and SURFTAB2 to 6.

5 Revolve the profiles about vertical lines.

6 Change the colour of the revolved chair to blue, the table being red.

7 Restore UCS BASE.

8 Polar array the chair for four items about the origin.

9 Hide, shade, save as R14MOD\GARDEN.

Chapter 19

Edge surface

An edge surface is a 3D polygon mesh stretched between four **touching** edges. The edges can be combinations of lines, arcs, polylines or splines but **must form a closed loop**.

The edge surface mesh is controlled by the system variables:
a) SURFTAB1: the M facets in the direction of the first edge selected
b) SURFTAB2: the N facets in the direction of the edges adjacent to the first selected edge.

Three examples will be used to demonstrate the command, the first being in 2D, the second to allow us to use the editing features of a polygon mesh, and the third will use splines as the four touching edges.

Example 1 – 2D edge surfaces

1 Open any 2D drawing and make two layers, EDGE colour red and MESH colour blue – refer to Fig. 19.1. Display toolbars as needed.

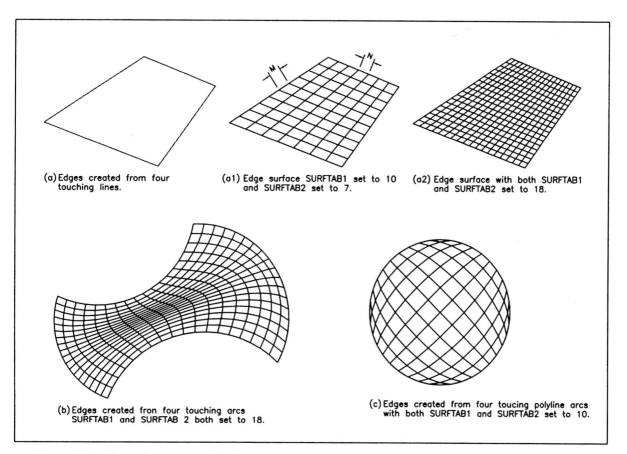

(a) Edges created from four touching lines.

(a1) Edge surface SURFTAB1 set to 10 and SURFTAB2 set to 7.

(a2) Edge surface with both SURFTAB1 and SURFTAB2 set to 18.

(b) Edges created fron four touching arcs SURFTAB1 and SURFTAB 2 both set to 18.

(c) Edges created from four toucing polyline arcs with both SURFTAB1 and SURFTAB2 set to 10.

Figure 19.1 Edge surface example 1 in 2D.

2 With layer EDGE current, create the following touching edges:
   a) four lines – fig. (a)
   b) four arcs – fig. (b)
   c) four single 90° polyline arcs – fig. (c).

3 Set SURFTAB1 to 10 and SURFTAB2 to 7.

4 Select the EDGE SURFACE icon from the Surfaces toolbar and:
   prompt   `Select edge 1` and pick a line
   prompt   `Select edge 2` and pick another line
   prompt   `Select edge 3` and pick a third line
   prompt   `Select edge 4` and pick the fourth line.

5 A 10 × 7 surface mesh is stretched between the four touching lines as fig. (a1).

6 *a*) Erase the added edge surface.
   b) Set SURFTAB1 and SURFTAB2 to 18.
   c) Menu bar with **Draw–Surfaces–Edge Surface** and pick the four touching lines in any order.
   d) The edge surface mesh is displayed as fig. (a2).

7 At the command line enter **EDGESURF <R>** and pick the four arcs to display the edge surface mesh as fig. (b).

8 Set both SURFTAB1 and SURFTAB2 to 10 and add an edge surface mesh between to four touching polyarcs – fig. (c). The result of this mesh is quite interesting?

9 This completes the first exercise and it need not be saved.

Example 2 – a 3D edge surface mesh

1 Open your MV3DSTD template file and refer to Fig. 19.2.

2 With layer MODEL, UCS BASE and the lower left viewport active, use the LINE icon and draw:
   From   0,0,0   *To*   150,0,–20   *To*   180,200,30
   To   40,120,50   *To*   close.

3 The four lines will be displayed as fig. (a).

4 Centre each viewport about the point 90,100,25 at 250 mag.

5 Make a new layer, MESH colour blue and current.

6 Set both SURFTAB1 and SURFTAB2 to 10.

7 Using the edge surface icon, pick the four lines in the order 1–2–3–4 as fig. (a).

8 An edge surface mesh will be stretched between the four lines as fig. (b).

9 In paper space, zoom the 3D viewport and return to model space.

10 Menu bar with **Modify–Object–Polyline** and:
   prompt   `Select polyline`
   respond   **pick any point on mesh**
   prompt   `Edit vertex/Smooth surface/...`
   enter   **E <R>** – the edit vertex option
   prompt   `Vertex (0,0). Next/Prev/...`
   and   an X is displayed at the 0,0 vertex – leftmost?
   enter   **U <R>** until X is at vertex (10,0), i.e.
   prompt   `Vertex (10,0). Next/Prev/...`
   enter   **M <R>** – the move option

prompt   Enter new location
enter   **@0,0,60 <R>**
and   **do not exit command.**

11 We now want to alter other vertices of the mesh to create a raised effect. This will be achieved by moving the X to the required vertices, using the M option and entering the required relative vertex coordinates.

12 Use the N/D/L/R/U options and enter the following new locations for the named vertices:

relative movement	*vertices*					
@0,0,50	9,0	10,1				
@0,0,40	8,0	9,1	10,2			
@0,0,30	7,0	8,1	9,2	10,3		
@0,0,20	6,0	7,1	8,2	9,3	10,4	
@0,0,10	5,0	6,1	7,2	8,3	9,4	10,5

1 3 When all the new vertex locations have been entered:
   a) enter X <R> to exit the edit vertex option
   b) then enter X <R> to end the command.

14 The mesh will be displayed with a 'raised corner' as fig. (c).

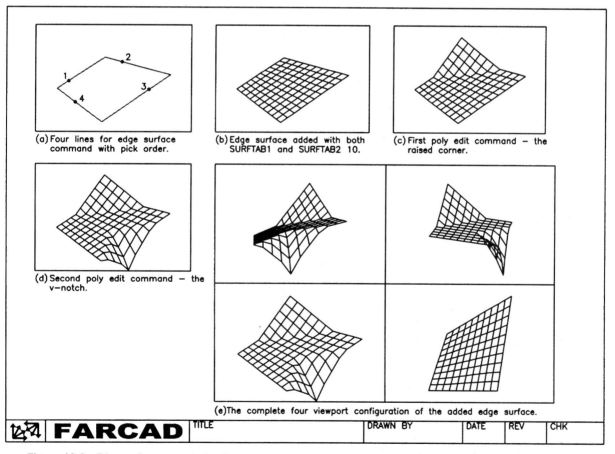

(a) Four lines for edge surface command with pick order.

(b) Edge surface added with both SURFTAB1 and SURFTAB2 10.

(c) First poly edit command – the raised corner.

(d) Second poly edit command – the v–notch.

(e) The complete four viewport configuration of the added edge surface.

FARCAD | TITLE | DRAWN BY | DATE | REV | CHK

Figure 19.2   Edge surface example 2 – 3D with edited vertices.

15 At the command line enter **PEDIT <R>** then:
   a) pick any point on the mesh
   b) enter E <R> for the edit vertex option
   c) use the N/U/R/L/D entries to move the X to the following named vertices and enter the following new locations (M entry):

relative movement	*vertices*						
@0,0,–80	4,10						
@0,0,–50	3,10	4,9	5,10				
@0,0,–30	2,10	3,9	4,8	5,9	6,10		
@0,0,–10	1,10	2,9	3,8	4,7	5,8	6,9	7,10

   d) exit the vertex option with X <R>
   e) exit the polyline edit command with X <R>.

16 These vertex modifications have produced a V-type notch in the mesh as fig. (d).

17 Paper space and zoom–previous then model space.

18 The complete four viewport configuration of the edge surface mesh is displayed in fig. (e).

19 The exercise is now complete – save if required.

20 *Investigate.*
   The effect of the smooth surface option (S) on the mesh.

Example 3 – an edge surface mesh created from splines

This example will demonstrate how an edge surface can be stretched between four spline curves to simulate a car body panel.

1 Open your MV3DSTD template file with layer MODEL, UCS BASE and the lower left viewport active. Refer to Fig. 19.3.

2 With the SPLINE icon from the Draw toolbar, draw four spline curves using the following coordinates, picking the start and end tangent points to suit your 'own panel design':

spline 1	*spline 2*	*spline 3*	*spline 4*
0,0,0	–200,0,0	–200,80,105	0,0,0
–200,0,0	–200,0,100	0,100,105	0,0,110
right-click	–200,80,105	right-click	0,100,105
	right-click		right-click

3 The four spline curves will be displayed as fig. (a).

4 Make a new layer, MESH colour blue and current.

5 Zoom centre about –100,50,50 at 200 magnification.

6 Set both SURFTAB1 and SURFTAB2 to 18.

7 Using the Edge Surface icon, pick the four spline curves to add the surface mesh.

8 Hide the model – any difference?

9 Save if required, as this completes the edge surface exercises.

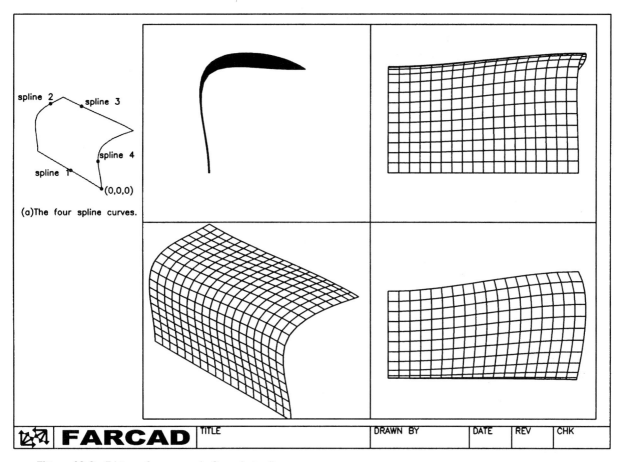

Figure 19.3 Edge surface example 3 – using splines.

Summary

1 An edge surface is a polygon mesh stretched between four **touching** objects – lines, arcs, splines or polylines.

2 An edge surface mesh can be edited with the polyline edit command.

3 The added surface is a **COONS** patch and is **bicubic**, i.e. one curve is defined in the mesh M direction and the other is defined in the mesh N direction.

4 The first curve (edge) selected determines the mesh M direction and the adjoining curves define the mesh N direction.

5 The mesh density is controlled by the system variables:
   a) SURFTAB1: in the mesh M direction
   b) SURFTAB2: in the mesh N direction.

6 The default value for SURFTAB1 and SURFTAB2 is 6.

7 The type of mesh stretched between the four curves is controlled by the **SURFTYPE** system variable and:
   a) SURFTYPE 5 – Quadratic B-spline
   b) SURFTYPE 6 – Cubic B-spline (default)
   c) SURFTYPE 8 – Bezier curve.

8   The SURFTYPE variable controls the appearance of all mesh curves.

9   Using different types of curves should be investigated by the user, as the default (6) is considered suitable for our purposes.

Assignment

The activity which I have set for you is very similar to the second example, i.e. an added edge surface has to have several of its vertices modified to give a 'flat-top hill' effect. The process is quite tedious, but persevere with it as it is needed for another activity in a later chapter.

Activity 13: a flat-topped hill

1   Use your MV3DSTD template file with UCS BASE.

2   Zoom centre about 0,0,50 at 400 magnification originally.

3   With layer MODEL, create four touching polyline arcs of radius 200 with 0,0 as the arc centre point.

4   Make a new layer called HILL, colour green and current.

5   Set SURFTAB1 and SURFTAB2 to 20.

6   A*d*d an edge surface to the four touching polyarcs.

7   Use the Edit Polyline command with the Edit Vertex option to move the following vertices by @**0,0,100**:

a)			6,9	6,10	6,11			
b)	7,7	7,8	7,9	7,10	7,11	7,12	7,13	
c)	8,7	8,8	8,9	8,10	8,11	8,12	8,13	
d) 9,6	9,7	9,8	9,9	9,10	9,11	9,12	9,13	9,14
e) 10,6	10,7	10,8	10,9	10,10	10,11	10,12	10,13	10,14
f) 11,6	11,7	11,8	11,9	11,10	11,11	11,12	11,13	11,14
g)	12,7	12,8	12,9	12,10	12,11	12,12	12,13	
h)	13,7	13,8	13,9	13,10	13,11	13,12	13,13	
i)			14,9	14,10	14,11			

8   *Note.*

  a) The named vertices all lie within a circle of radius 100.
  b) Use the N/U/D/L/R entries of the edit vertex option until the named vertex is displayed then use the M option with an entry of @0,0,100.

9   When all the vertices have been modified, optimize your viewpoints. I used four different VPOINT ROTATE values and the effect with hide was quite 'pleasing'.

10 Save this drawing as **R14MOD\HILL** as it will be used with a later activity.

Chapter 20

3D polyline

A 3D polyline is a continuous object created in 3D space. It is similar to a 2D polyline, but does not possess the 2D versatility, i.e. there are no variable width or arc options available with a 3D polyline.

Example

This exercise will create a series of hill contours from splined 3D polylines, so:

1 Open your MV3DSTD template file with the lower left viewport active and UCS BASE. Refer to Fig. 20.1.

2 Make the following new layers, LEVEL1, LEVEL2, LEVEL3, LEVEL4, PATH1 and PATH2 the colours to your own specification.

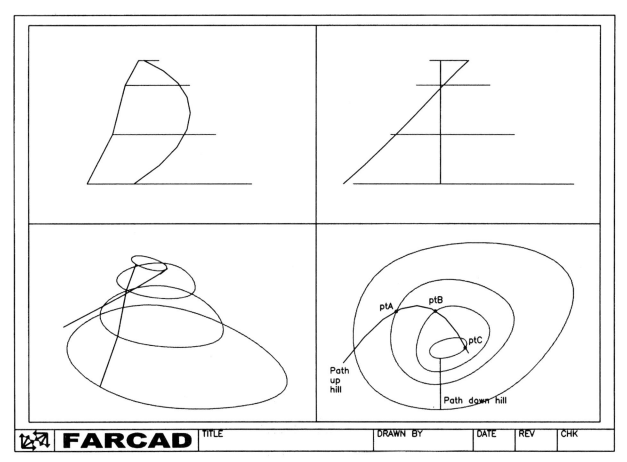

Figure 20.1 3D polyline example.

3 With layer LEVEL1 current, menu bar with **Draw–3Dpolyline** and:
   prompt   From point and enter: **0,50,0 <R>**
   prompt   Close/Undo/<Endpoint of line> and enter: **150,0,0 <R>**
   prompt   Endpoint of line and enter: **220,60,0 <R>**
   prompt   Endpoint of line and enter: **260,140,0 <R>**
   prompt   Endpoint of line and enter: **170,180,0 <R>**
   prompt   Endpoint of line and enter: **20,160,0 <R>**
   prompt   Endpoint of line and enter: **c <R>**.

4 Make layer LEVEL2 current and at the command line enter **3DPOLY <R>** and:
   prompt   From point and enter: **40,60,50 <R>**
   prompt   Endpoint of line and enter: **80,20,50 <R>**
   prompt   Endpoint of line and enter: **140,35,50 <R>**
   prompt   Endpoint of line and enter: **200,100,50 <R>**
   prompt   Endpoint of line and enter: **130,140,50 <R>**
   prompt   Endpoint of line and enter: **60,130,50 <R>**
   prompt   Endpoint of line and enter: **c <R>**.

5 With layer LEVEL3 current, use the 3D polyline command to enter the following coordinate values:
   From   70,70,100   *Endpoint* 80,35,100   *Endpoint* 130,45,100
   Endpoint 170,90,100   *Endpoint* 100,120,100   *Endpoint* close.

6 With layer LEVEL4 current, enter the following 3D polyline coordinates:
   From   85,70,125   *Endpoint* 90,50,125   *Endpoint* 130,60,125
   Endpoint 130,80,125   *Endpoint* close.

7 Menu bar with **Modify–Object–Polyline** and:
   prompt   Select polyline
   respond   **pick the 3D polyline created at level 1**
   prompt   Open/Edit vertex/...
   enter   **S <R>** – the spline option
   prompt   Open/Edit vertex/...
   enter   **X <R>** – the exit command option.

8 The selected polyline will be displayed as a splined curve.

9 Use the S option of the Edit Polyline command to spline the other three 3D polylines.

10 Zoom centre about 120,90,60 at 200 magnification.

11 With layer PATH1 current, use the 3D polyline command with the following entries:
   From   0,50,0 – level 1 point
   Endpoint 60,130,50 – level 2 point
   Endpoint 100,120,100 – level 3 point
   Endpoint 130,60,125 – level 4 point
   Endpoint <R>.

12 This 3D polyline is a path 'up the hill' and each entered coordinate value is a point on the contours.

13 Spline this polyline – does not pass through the entered values?

14 *Task*.
    a) With layer PATH2 current, create a 3D polyline path 'down the hill' giving the shortest distance path. The polyline must touch each contour line.
    b) Using the ID command, identify the coordinates of points A, B and C. My values were:
       ptA: 55.06, 101.57, 50
       ptB: 95.42, 101.91, 100
       ptC: 125.52, 63,61, 125.
    c) Find the value of the shortest distance down path2. My value was 136.40.

15 The exercise is now complete. Save if required.

Summary

1 A 3D polyline is a single object and can be used with an *x*, *y*, *z* coordinate entry.

2 A 3D polyline can be edited with options of Edit vertex, Spline and decurve.

3 3D polylines cannot be displayed with varying width or with arc segments.

Chapter 21

3D objects

AutoCAD has nine predefined 3D objects, these being box, pyramid, wedge, dome, sphere, cone (and cylinder), torus, dish and mesh. The are 'real' 3D objects, that is, they can be displayed with hidden line removal, using the HIDE command.

To demonstrate some of these objects:

1 Open your MV3DSTD template file, with layer MODEL, UCS BASE and the lower left viewport active. Refer to Fig. 21.1 and display the Surfaces toolbar.

2 In the steps which follow, reason out the coordinate entry values.

3 Select the BOX icon from the Surfaces toolbar and:
   prompt   Corner of box and enter: **0,0,0 <R>**
   prompt   Length and enter: **150 <R>**
   prompt   Cube/<Width> and enter: **120 <R>** – width option
   prompt   Height and enter: **100 <R>**
   prompt   Rotation angle about Z axis and enter: **0 <R>**.

4 A red box will be displayed as the 0,0,0 origin point.

5 Menu bar with **Draw–Surfaces–3D Surfaces** and:
   prompt   3D Objects dialogue box
   respond   **pick Wedge then OK**
   prompt   Corner of wedge and enter: **150,0,0 <R>**
   prompt   Length and enter: **80 <R>**
   prompt   Width and enter: **70 <R>**
   prompt   Height and enter: **150 <R>**
   prompt   Rotation angle about Z axis and enter: **–10 <R>**.

6 Using the PROPERTIES icon from the Object Properties toolbar, change the colour of the wedge to yellow.

7 Using the icons from the Surfaces toolbar, or the 3D Objects dialogue box, create the following two 3D objects:
   Cone
   Base centre point: 50,70,100
   Radius of base: 50
   Radius of top: 0
   Height: 85
   Number of segments: 16
   Colour: green

   Cylinder – using cone
   Base centre point: 75,0,50
   Radius of base: 50
   Radius of top: 50
   Height: 90
   Number of segments: 16
   Colour: blue.

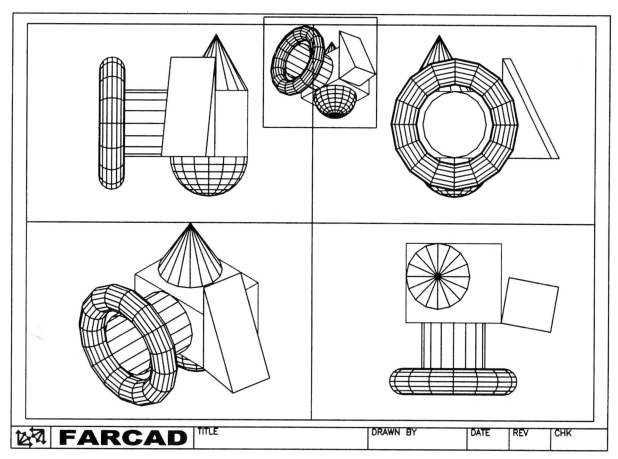

Figure 21.1 3D object example.

8 Restore UCS RIGHT and with the ROTATE icon from the Modify toolbar:
   a) pick the blue cylinder then right-click
   b) base point: 0,50
   c) angle: 90
   d) restore UCS BASE.

9 Create the following two 3D objects:
   Dish
   Centre of dish: 75,60,0
   Radius: 60
   Number of longitudinal segments: 16
   Number of latitudinal segments: 16
   Colour: magenta.

Torus
Centre of torus: 75,–90,50
Radius of torus: 100
Radius of tube: 20
Segments around tube circumference: 16
Segments around torus circumference: 16
Colour: cyan.

10   Restore UCS RIGHT and with the ROTATE icon:
   a) pick the cyan torus then right-click
   b) base point: –90,50
   c) angle: 90
   d) restore UCS BASE.

11   Zoom centre about 75,0,50 at 300 magnification.

12   Hide and shade to display the model with 'bright' colours.

13   *Task*.
   a) Enter paper space and make layer VP current.
   b) Menu bar with View–Floating Viewports–1 Viewport and make a new viewport selecting one of the points 'outside' the original four as shown in Fig. 21.1.
   c) Display the model in this new viewport from below and zoom centre about the same point as before.
   d) *Question*: why did we pick one of the new viewport points outside the original viewports?

14   Save the layout, however we will not use it again.

Summary

1   The nine 3D objects are displayed as either faced or meshed surface models.

2   The hide and shade (and render) commands can be used with 3D objects.

3   3D objects are created from a reference point (corner/centre, etc.) and certain model geometry, e.g. length, width and height.

4   A cylinder is created from a cone with the base and top radii the same.

5   Activating 3D objects is available from the Surfaces toolbar or from the 3D Objects dialogue box. The various objects can also be created by command line entry, although this is **not** recommended.

Assignment

Macfaramus (the famous Egyptian builder) was commissioned by King Tootencadum to design and build a palace for queen Nefersaydy. With his knowledge of CAD, Macfaramus decided to build the palace from 3D objects, and this is your assignment.

Activity 14: palace

1   Use your MV3DSTD template file with UCS BASE.

2   The 3D objects have to be positioned in a circle with an 85 maximum radius. This circle has to be:
   a) created from four touching polyarcs – as a previous exercise
   b) edge surfaced with both SURFTAB1 and SURFTAB2 set to 16. The edge surface mesh should be on its own layer with a colour number of 42.

3   The 3D objects have to be created on layer MODEL and can be to your own specification and layout. Some of my 3D objects were:

box	*wedge*	*dome*
corner: −40,−40,0	corner: 40,−40,0	centre: 0,0,60
length: 80	length: 20	radius: 25
width: 80	width: 10	colour: magenta
height: 60	height: 60	
colour: red	colour: blue	
	copied and rotated	

cylinder	*cone*
centre: 50,−50,0	centre: 50,−50,70
radius: 8	radius: 12
height: 70	height: 20
colour: green	colour: green
copied	copied.

4   When the palace layout is complete, hide and shade.

5   Save the complete model as **R14MOD\PALACE**. It will be used in a later exercise.

Chapter 22

3D geometry commands

All AutoCAD commands can be used in 3D but there are three commands which are specific to 3D models, these being 3D Array, Mirror 3D and Rotate 3D. In this chapter we will investigate these three commands as well as how to use the ALIGN command with 3D models.

Getting started

To investigate the 3D commands, we will create a new model from 3D Objects, so:

1 Begin a new drawing with your MV3DSTD template file, with layer MODEL, UCS BASE and the lower left viewport active. Refer to Fig. 22.1.

2 Select the BOX icon from the Surfaces toolbar and:
   a) corner: 0,0,0
   b) length: 100; width: 100; height: 40; Z rotation: 0
   c) colour: red.

3 Select the WEDGE icon from the Surfaces toolbar and:
   a) corner: 0,0,40
   b) length: 30; width: 30; height: 30; Z rotation: 0
   c) colour: blue.

4 In each viewport, zoom centre about 50,50,35 at 200 magnification.

5 The two 3D objects will be displayed as fig. (a).

Rotate 3D

Using the menu bar sequence Modify–Rotate or selecting the ROTATE icon from the Modify toolbar results in a 2D command, i.e. the selected objects are rotated in the current *XY* plane. Objects can be rotated in 3D relative to the *X*-, *Y*- and *Z*-axes with the Rotate 3D command. The command will be demonstrated by:
a) rotating the blue wedge
b) rotating the complete model.

Rotating the wedge

1 From the menu bar select **Modify–3D Operation–Rotate 3D** and:

prompt	Select objects
respond	**pick the blue wedge then right-click**
prompt	Axis by Object/Last/...
enter	**X <R>** – the *X*-axis option
prompt	Point on X axis<0,0,0>
enter	**0,0,40 <R>** – why these coordinates?
prompt	<Rotation angle>/Reference
enter	**90 <R>**.

2   The blue wedge is rotated about the *X*-axis as Fig. 22.1(b).

3   Activate the Rotate 3D command and:

prompt	`Select objects`
enter	**P <R><R>** – the previous selection set option
prompt	`Rotate 3D options`
enter	**Z <R>** – the Z-axis option
prompt	`Point on Z axis<0,0,0>`
enter	**90 <R>**.

4   The blue wedge is now aligned as required – fig. (c).

5   Select the ARRAY icon from the Modify toolbar and:
    a) pick the blue wedge the right-click
    b) enter **P <R>** for the polar array option
    c) enter the centre point as 50,50
    d) number of items: 4
    e) angle to fill of 360 with rotation.

6   The blue wedge is arrayed to the four corners of the box – fig. (d).

7   Hide the model – fig. (e), shade and regen.

8   Save at this stage as **R14MOD\3DGEOM** for future recall.

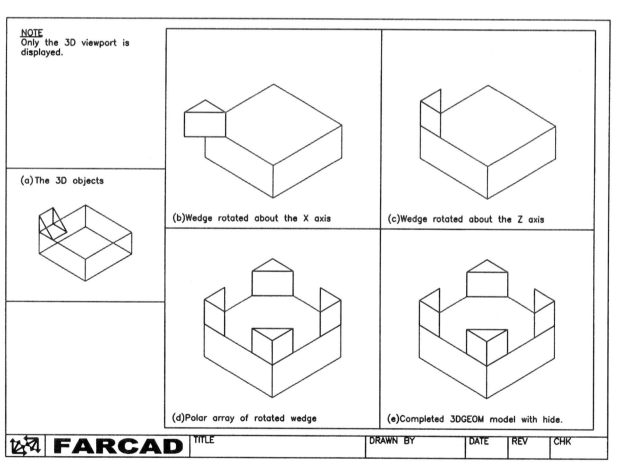

Note: Only the 3D viewport is displayed.

(a) The 3D objects

(b) Wedge rotated about the X axis

(c) Wedge rotated about the Z axis

(d) Polar array of rotated wedge

(e) Completed 3DGEOM model with hide.

FARCAD TITLE   DRAWN BY   DATE   REV   CHK

Figure 22.1 The 3DGEOM model for the 3D commands.

Rotating the model

1 Model 3DGEOM on screen with lower left viewport active, UCS BASE and layer MODEL current. Refer to Fig. 22.2.

2 Menu bar with Modify–3D Operation–Rotate 3D and:

prompt	Select objects
respond	**window the complete model then right-click**
prompt	Axis by Object/Last/...
enter	**X <R>** – the *X*-axis option
prompt	Point on X axis<0,0,0>
respond	**right-click**, i.e. accept the 0,0,0 default point
prompt	<Rotation angle>/Reference
enter	**90 <R>**
then	PAN to suit – fig. (b).

3 At the command line enter **ROTATE3D <R>** and:

prompt	Select objects
respond	**window the model then right-click**
prompt	Axis by Object/...
enter	**Z <R>** – the *Z*-axis option
prompt	Point on Z axis<0,0,0> and: right-click
prompt	<Rotation angle>/Reference
enter	**90 <R>**
then	PAN to suit – fig. (c).

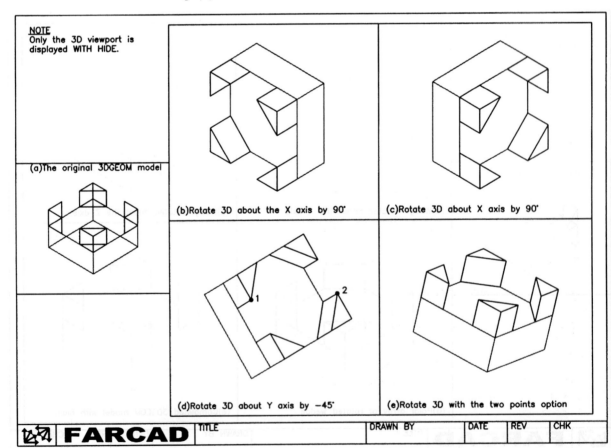

NOTE
Only the 3D viewport is displayed WITH HIDE.

(a)The original 3DGEOM model

(b)Rotate 3D about the X axis by 90°

(c)Rotate 3D about X axis by 90°

(d)Rotate 3D about Y axis by −45°

(e)Rotate 3D with the two points option

FARCAD | TITLE | DRAWN BY | DATE | REV | CHK

Figure 22.2 Rotate 3D command using 3DGEOM model.

4 Activate the Rotate 3D command and:
   a) window the model
   b) select the *Y*-axis option
   c) accept the default 0,0,0 point
   d) enter –45 as the rotation angle, pan to suit – fig. (d).

5 Rotate 3D again, window the model and:
   prompt Axis by Object/.../<2 points>
   respond **Intersection icon and pick pt1**
   prompt 2nd point on axis
   respond **Intersection icon and pick pt2**
   prompt <Rotation angle>/Reference
   enter   **–45 <R>**
   then   PAN to suit – fig. (e).

6 Hide the viewports.

7 Save layout if required – we will not use it again.

Mirror 3D

This command allows objects to be mirrored about selected points or about any of the three *X–Y–Z* planes.

1 Open model 3DGEOM, UCS BASE, lower left viewport active and refer to Fig. 22.3.

2 Menu bar with **Modify–3D Operation–Mirror 3D** and:
   prompt Select objects
   respond **window the model then right-click**
   prompt Plane by Object/Last/Zaxis/...
   enter   **XY <R>** – the *XY* plane option
   prompt Point on plane<0,0,0> and: **right click**
   prompt Delete old objects<N> and enter: **Y <R>**
   then   PAN to suit – fig. (b).
   Note: the model at this stage has the ambiguity effect of all 3D models, i.e. are you looking down or looking up?

3 At the command line enter **MIRROR3D <R>** and:
   prompt Select objects
   respond **window the model then right-click**
   prompt Axis by Object/...
   respond **Intersection icon and pick pt1**
   prompt 2nd point on plane
   respond **Intersection icon and pick pt2**
   prompt 3rd point on plane
   respond **Intersection icon and pick pt3**
   prompt Delete old object<N> and enter: **Y <R>**
   then   PAN to suit – fig. (c).

4 Activate the Mirror 3D command and:
   a) window the model
   b) select the *YZ* plane option
   c) pick point A as a point on the plane
   d) enter Y to delete old objects prompt
   e) pan to suit – fig. (d).

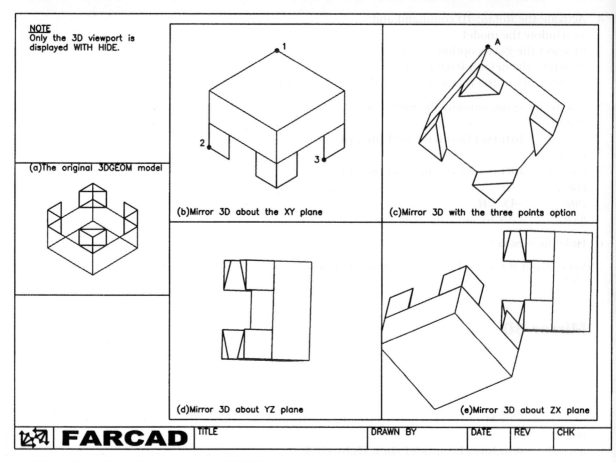

Figure 22.3 Mirror 3D command using 3DGEOM model.

5 Using the Mirror 3D command:
  a) window the model
  b) select the *ZX* option
  c) accept the default 0,0,0 point on plane
  d) accept the N default delete option
  e) pan as required – fig. (e).

6 Hide the viewports then save if required.

3D array

The 3D array command is similar in operation to the 2D array. Both rectangular and polar arrays are possible, the rectangular array having rows and columns as well as levels in the Z-direction. The result of a 3D polar array requires some thought!

Rectangular

1 Open 3DGEOM, UCS BASE, lower left active.

2 Menu bar with **Modify–3D Operation–3D Array** and:

prompt	Select objects
respond	**window the model then right-click**
prompt	Rectangular or Polar array and enter: **R <R>**
prompt	Number of rows(--)<1> and enter: **2 <R>**
prompt	Number of columns(||||)<1> and enter: **3 <R>**
prompt	Number of levels(…)<1> and enter: **4 <R>**
prompt	Distance between rows(--) and enter: **120 <R>**
prompt	Distance between columns(||||) and enter: **120 <R>**
prompt	Distance between levels(…) and enter: **100 <R>**.

3 The model will be displayed in a 2 × 3 × 4 rectangular matrix pattern but will probably need to be re-centred. Zoom centre about the point 170,110,185 (why these coordinates?) at 400 magnification but 550 in the 3D viewport.

4 Hide the model – Fig. 22.4 and shade?

5 This exercise does not need to be saved.

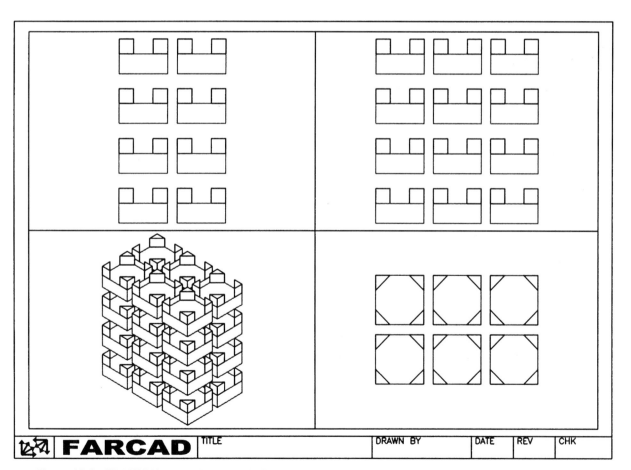

Figure 22.4 3D ARRAY (rectangular) using 3DGEOM model.

Polar

1 Open 3DGEOM, UCS BASE and lower left viewport active and draw a line from −50,0,0 to @0,0,100.

2 At the command line enter **3DARRAY <R>** and:

prompt	Select objects
respond	**window the model then right-click**
prompt	Rectangular or polar and enter: **P <R>**
prompt	Number of items and enter: **5 <R>**
prompt	Angle to fill and enter: **360 <R>**
prompt	Rotate objects and enter: **Y <R>**
prompt	Center point of array
respond	**Endpoint icon and pick one end of the line**
prompt	Second point on axis of rotation
respond	**Endpoint icon and pick the other end of the line**.

3 Zoom all and hide each viewport – fig. (a).

4 Undo the polar array and erase the line.

5 Draw a new line from −50,0,0 to @0,100,0.

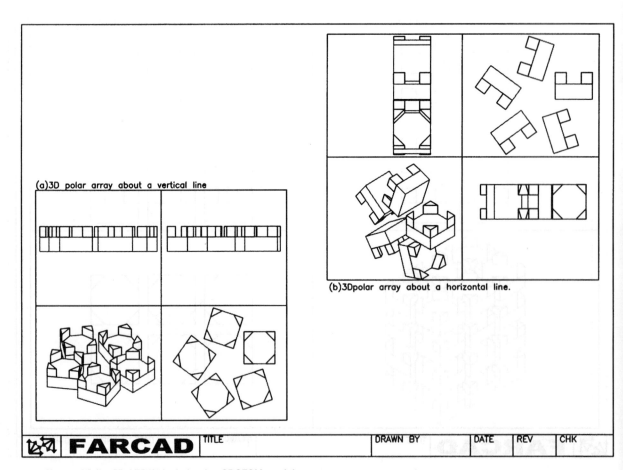

(a)3D polar array about a vertical line

(b)3Dpolar array about a horizontal line.

FARCAD | TITLE | | DRAWN BY | DATE | REV | CHK |

Figure 22.5 3D ARRAY (polar) using 3DGEOM model.

6 Activate the polar array command, window the model and:
   a) number of items: 5
   b) angle to fill: 360
   c) rotate as copied: Y
   d) centre point of array: one endpoint of line
   e) second point on axis: other endpoint of line
   f) zoom all and hide – fig. (b).

7 This exercise is complete and need not be saved.

Align

The align command can be used in 2D or 3D and allows objects (models) to be aligned with each other.

1 Open your MV3DSTD template file and create a single 3D viewport with UCS BASE and layer MODEL current. Refer to Fig. 22.6.

2 Using the Surfaces toolbar create the following objects:

	Box	Wedge
corner	0,0,0	160,0,0
length	100	100
wedge	80	80
height	50	50
rotation	0	0
colour	red	green

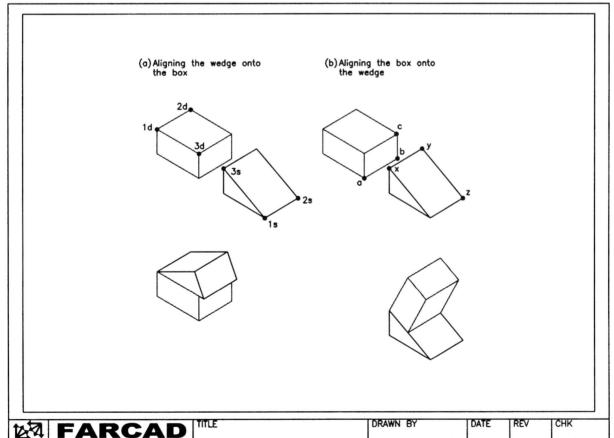

Figure 22.6 The ALIGN command with 3D objects.

3 Copy the box and wedge from 0,0,0 to @200,200.

4 Menu bar with **Modify–3D Operation–Align** and:

prompt	Select objects
respond	**pick the green wedge then right-click**
prompt	Specify 1st source point
respond	**Intersection icon and pick pt1s**
prompt	Specify 1st destination point
respond	**Intersection icon and pick pt1d**
prompt	Specify 2nd source point
respond	**Intersection icon and pick pt2s**
prompt	Specify 2nd destination point
respond	**Intersection icon and pick pt2d**
prompt	Specify 3rd source point
respond	**Intersection icon and pick pt3s**
prompt	Specify 3rd destination point
respond	**Intersection icon and pick pt3d**.

5 The green wedge will be aligned with its sloped surface on the top of the box – fig. (a).

6 At the command line enter **ALIGN <R>** and:

   a) pick the red box then right-click
   b) pick 1st points a and x
   c) pick 2nd points b and y
   d) pick 3rd points c and z
   e) the red box will be aligned onto the sloped surface of the wedge – fig. (b).

7 This completes the align exercise.

3D extend and trim

Release 14 allows 3D objects to be extended and trimmed to an *XY* plane. The two commands are very dependent on the UCS position.

1 Open your MV3DSTD template file and refer to Fig. 22.7.

2 Draw a square using the LINE icon:

   from: 0,0 to: @100,0 to: @0,100 to: @–100,0 to: close.

3 Multiple copy the square from: 0,0 to: @0,0,120 and to: @0,0,180.

4 Scale the top square about the point 50,50,180 by 0.5.

5 Draw in the following four lines:

from	*to*	*colour*
a) 25,25,180	@0,–5,–30	blue
b) 75,25,180	@5,0,–30	green
c) 75,75,180	@0,5,–30	cyan
d) 25,75,180	@0,0,–60	magenta

6 Draw in lines aw, bx, cy and dz to complete the model as fig. (a).

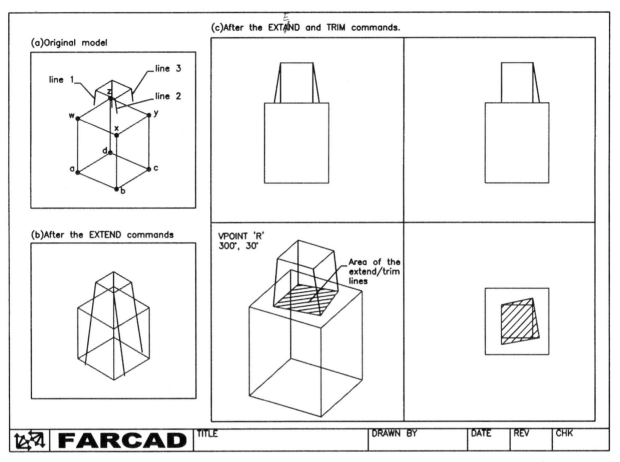

Figure 22.7 EXTEND and TRIM in 3D.

Extend

1 Restore UCS RIGHT and select the EXTEND icon from the Modify toolbar and:

prompt Select objects

respond **pick line ab then right-click**

prompt <Select object to extend>/Project/Edge...

enter **E <R>** – the edge option

prompt Extend/No extend...

enter **E <R>** – the extend option

prompt <Select object to extend>/Project/Edge...

enter **P <R>** – the project option

prompt None/Ucs/View...

enter **U <R>** – the current UCS option

prompt <Select object to extend>...

respond **pick blue line 1 then right-click**.

2 Repeat the EXTEND command using the entries E,E,P,U as step (1) and:

   a) extend the green line 2 to edge bc with UCS RIGHT

   b) extend the cyan line 3 to edge cd with UCS FRONT.

3 The extended lines are displayed as fig. (b).

Trim

1 Restore UCS BASE, select the TRIM icon and:
   prompt   Select objects
   respond   **pick line wx then right-click**
   prompt   <Select object to trim>/Project/Edge...
   enter   **P <R>** – the project option
   prompt   None/Ucs/View...
   enter   **V <R>** – the view option
   prompt   <Select object to trim>
   respond   *a*) make the top right viewport active
               b) pick the blue line then right-click.

2 Repeat the TRIM command with P and V entries as step (1) and:
   a) trim the green line to edge xy, picking the green line in the top left viewport at the select object to trim prompt
   b) trim the cyan line to edge yz, picking the cyan line in the top right viewport at the select object to trim prompt.

3 The coloured lines have now been extended and trimmed 'to the top surface' of the large red box – fig. (c).

4 *Task*.
   a) draw four lines connecting the 'bottom ends' of the blue, green, cyan and magenta lines
   b) hatch this area
   c) find the hatched area and perimeter. My values were: Area: 3350 Perimeter: 232.65.

Summary

1 The commands Rotate3D, Mirror3D and 3Darray are specific to 3D models.

2 Rotate 3D allows models to be rotated about the *X*-, *Y*- and *Z*-axes as well as two specified points and about objects.

3 Mirror 3D allows models to be mirrored about the *XY*, *YZ* and *ZX* axes as well as three specified points and objects.

4 3D Array is similar to the 2D command but has 'levels' in the *Z* direction.

5 3D models can be aligned with each other.

Chapter 23

Blocks and Wblocks in 3D

3D blocks and wblocks are created and inserted into a drawing in a similar manner as 2D blocks and wblocks. The UCS position and orientation is critical. In this chapter we will:
a) create a chess set using blocks
b) create a wall clock as wblocks.

Creating the block models

1 Open your MV3DSTD template file, layer MODEL, UCS BASE and lower left viewport active. Refer to Fig. 23.1.

2 Create the following two 3D box objects:

	box1	box2
corner	0,0,0	0,120,0
length	80	80
width	80	80
height	10	10
rotation	0	0
colour	magenta	number 30

3 Restore UCS FRONT and make the upper right viewport active. Draw a line, from 150,–10 to @0,100.

4 *a*) Draw the pawn outline as a polyline from: 150,0 using your own design with the sizes given as reference
   b) set SURFTAB1 to 16
   c) with the REVOLVED SURFACE icon from the Surfaces toolbar, revolve the pawn outline about the vertical line – full circle
   d) the pawn colour is to be red.

5 Still with UCS FRONT, draw a line from 210,–10,–60 to @0,120.

6 *a*) Draw the rook (castle) outline as a polyline from: 210,0,–60 using the reference sizes given but your own design
   b) revolve the rook polyline about the vertical line for a full circle
   c) the rook colour is to be red.

7 Erase the two vertical lines.

8 Restore UCS BASE and zoom centre about 120,90,50 at 250 mag to display the models as Fig. 23.1.

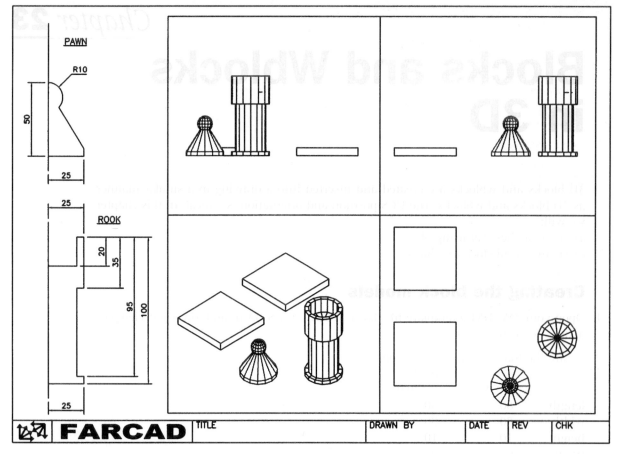

Figure 23.1 3D block details.

Making the blocks

1 Make the lower left viewport active.

2 At the command line enter **BLOCK <R>** and:

prompt	Block name(or ?)
enter	**SQ1 <R>**
prompt	Insertion base point and enter: **0,0,0 <R>**
prompt	Select objects
respond	**pick the magenta box then right-click**
and	the selected object disappears. Remember OOPS?

3 Repeat the command line BLOCK entry and:

prompt	Block name and enter: **PAWN <R>**
prompt	Insertion base point and enter: **150,0,0 <R>**
prompt	Select objects
respond	**pick the red pawn then right-click**
and	pawn disappears and original outline displayed.

4 Use the BLOCK command twice more with:
   a) block name: SQ2
      insertion base point: 0,120,0
      select objects: pick orange box then right-click.
   b) block name: ROOK
      insertion base point: 210,60,0
      select objects: pick red rook then right-click
      and original outline is displayed.

5 Erase the two original polyline outlines.

6 *Note*: blocks can be made using a dialogue box with the menu bar sequence **Draw–Block–Make**.

Inserting the blocks

1 Four 'blank' viewports displayed?

2 At command line enter **BLOCK <R>** and:
prompt	Block name (or ?) and enter: **? <R>**
prompt	Block(s) to list<*> and right-click
and	AutoCAD Text Window with User-Defined blocks:
	PAWN ROOK SQ1 SQ2
respond	cancel the text window.

3 At the command line enter **INSERT <R>** and:
prompt	Block name (or ?) and enter: **SQ1 <R>**
prompt	Insertion point and enter: **0,0,0 <R>**
prompt	X scale factor and enter: **1 <R>**
prompt	Y scale factor and right-click, i.e. X=Y=1
prompt	Rotation angle and enter: **0 <R>**
and	the magenta box is inserted at the selected point.

4 Repeat the INSERT command and:
   Block name: SQ2
   Insertion point: 80,0,0
   Scale: X=1, Y=1, Rotation=0.

5 Zoom centre about 320,320,100 at 700 in all viewports.

6 Now complete the 64 square chess board using one of the following methods:
   a) inserting each block SQ1 and SQ2
   b) multiple copy the two inserted boxes
   c) copy and array.

7 When the 64 squares have been created, INSERT the following blocks, all full-size with 0 rotation:

name:	PAWN	ROOK	ROOK
insertion point:	40,120,10	40,40,10	600,40,10.

8 Rectangular array the red pawn with:
   a) rows: 1
   b) columns: 8
   c) column distance: 80.

9 Copy the two red rooks from: 40,40,10, to: @0,560,0.

10 Copy the eight red pawns from: 40,80,10, to: @0,400,0.

11 The layout should be as Fig. 23.2.

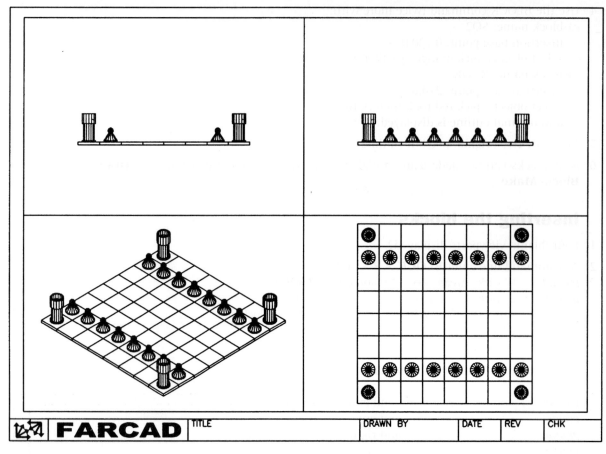

Figure 23.2 Inserting the blocks.

12 *Task*.
   a) Change the colour of the copied pawns and rooks to blue
   b) Problems – think inserted blocks!

13 Save the chess set at this stage as **R14MOD\CHESS**.

Creating the Wblock models

1 Open the MV3DSTD template file and zoom centre about 150,0,100 at 300 magnification in all viewports.

2 Restore UCS FRONT, lower left viewport active, layer MODEL and refer to Fig. 23.3.

3 Create the three outlines for parts of a wall clock with:
   a) draw the body as lines using the 0,0 start point and the reference sizes given. Use the polyline edit (PEDIT) command to 'convert' the five lines into a single polyline object
   b) draw the face as a circumscribed octagon with centre at 90,90,0 and radius 30
   c) draw the dial as a 30 radius circle, centre at 90,10,0.

4 Draw the following three lines:
   a) line 1, from: 150,0   to: @0,0,40
   b) line 2, from: 200,0   to: @0,0,15
   c) line 3, from: 250,0   to: @0,0,8.

5 The three wall clock components will be created as tabulated surface models, so set SURFTAB1 to 16.

6 With the TABULATED SURFACE icon from the Surfaces toolbar:

   prompt   Select path curve

   respond   **pick the BODY polyline**

   prompt   Select direction vector

   respond   **pick line 1 at end indicated**

   and   extruded red surface model of wall clock body.

7 Repeat the tabulated surface command and:

   a) select the FACE octagon as the path curve and line 2 as the direction vector

   b) select the DIAL circle as the path curve and line 3 as the direction vector.

   Note: remember to select the line direction vectors at the end indicated.

8 Change the colour of the extruded surface models:

   a) FACE: blue

   b) DIAL: green.

9 The layout should resemble Fig. 23.3.

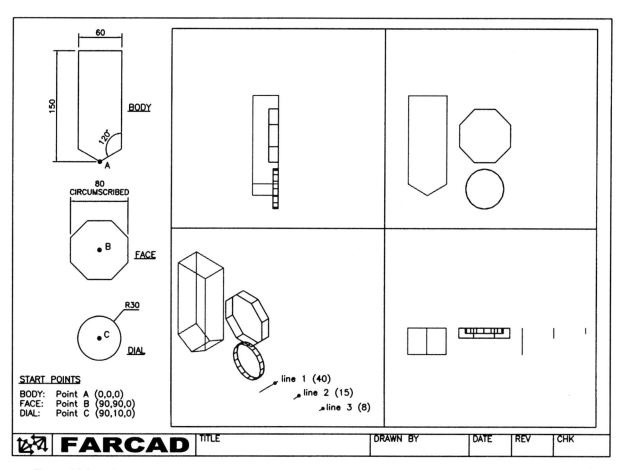

Figure 23.3 Information for creating the three wblocks.

Making the Wblocks

1 Restore the WCS, i.e. menu bar with **Tools–UCS–World**.

2 At the command line enter **WBLOCK <R>** and:
   prompt   Create Drawing File dialogue box
   respond   1. ensure r14mod is current folder
               2. ensure drawing type is *.dwg
               3. enter File name as: BODY
               4. pick Save
   prompt   Block name and right-click
   prompt   Insertion base point and enter: **50,50,0 <R>**
   prompt   Select objects
   respond   **pick the red extruded body then right-click**
   and   original outline displayed?

3 Create another two wblocks using the same procedure as step (2) with the following information:

	wblock1	*wblock2*
File name	FACE	DIAL
Block name	right-click	right-click
Insertion base point	140,50,90	140,50,10
Objects	blue octagon	green cylinder

Inserting the three wblocks

1 Menu bar with **File–New** ('No' to save changes) and open your MV3DSTD template file with UCS BASE, layer MODEL and the lower left viewport active.

2 Menu bar with **Insert–Block** and:
   prompt   Insert dialogue box
   respond   1. pick File
               2. pick R14MOD folder – C:\ needed?
               3. pick BODY
               4. pick Open
   prompt   Insert dialogue box with:
               a) Block: BODY
               b) File: C:\r14mod\BODY.dwg
   respond   **pick OK**
   prompt   Insertion point and enter: **0,0,0 <R>**
   prompt   Scale and enter X=1, Y=1, rotation = 0.

3 Repeat the Insert–Block menu bar selection and:
   a) pick the FACE file
   b) insertion point: **0,–40,130**
   c) full size with no rotation.

4 Insert the DIAL wblock, full size with zero rotation at the point **0,–55,130**.

5 Zoom centre about the point 0,–30,80 at 200 magnification in all viewports – Fig. 23.4.

6 Hide and shade the model – no 'tops'.

7 Save as this exercise is now complete.

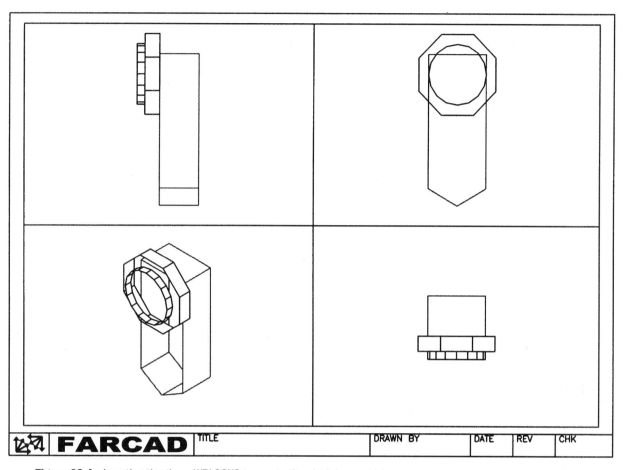

Figure 23.4 Inserting the three WBLOCKS to create the clock 'assembly'.

Summary

1 3D blocks and wblocks are created and inserted in a similar manner to 2D blocks and wblocks.

2 With 3D blocks, the position of the UCS is important.

3 With 3D wblocks, the WCS should be restore when creating the wblocks.

4 It is recommended that wblocks are 'stored' in the same folder as drawings.

Assignments

Two activities have been included for you to attempt, one with blocks using the chess set partially completed, and the other using wblocks with two previously saved drawings.

Activity 15: Chess set

1 Recall the drawing CHESS saved earlier in this chapter to display the 64 square chess board with the two sets of red and blue pawns and rooks.

2 Design the other chess pieces – KNIGHT, BISHOP, KING and QUEEN using the same method as the worked example:
   a) draw the outline as a polyline
   b) use the revolved surface command to create the piece as a 3D surface model
   c) *note*: the actual shape of the pieces is at your discretion. The information given in the Activity 15 drawing is for reference only
   d) ensure that your start point for the outline is known – it will be useful as the block insertion point.

3 Create a block of each created piece.

4 Insert the created blocks on the chess board.

5 Complete the chess set layout, remembering to change the colours of the pieces to red and blue as appropriate.

6 Hide and shade the model. Save as R14MOD\CHESS.

Activity 16: Palace of Queen Nefersaydy built by Macfaramus

Macfaramus was last encountered building the palace for Queen Nefersaydy. Unfortunately this palace was to be built on a flat-topped hill and you have to create the effect using wblocks.

1 Open the drawing R14MOD\HILL of the edge surface model created as Activity 13.

2 Insert the wblock drawing file R14MOD\PALACE of the 3D objects created as Activity 14.

3 The palace has to be inserted on the centre point of the hill top, and the coordinates of this point as 0,0,100. This is the only help given.

4 Optimize the viewport layout for maximum effect.

5 When complete save the layout as R14MOD\HILLPAL.

Chapter 24

3D surface model exercise

You now have the ability and experience to create 3D surface models. As a final 'test' I have included an exercise which requires several of the 3D surface commands to be used. This exercise has been used with students on the HNC/HND Computer Aided Draughting and Design courses at Motherwell College.

To complete the GLOBE exercise:

1 Open your MV3DSTD template file.

2 Refer to Fig. 24.1 which displays the globe in a traditional orthographic layout with all relevant sizes.

Note: as an additional exercise can you draw the three given views? What about an isometric view?

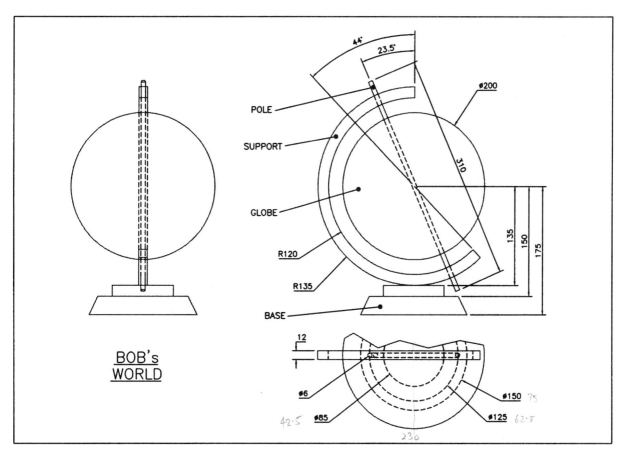

Figure 24.1 Globe reference sizes and layout.

3 The model can be considered as having four distinct 'parts' which I have named BASE, SUPPORT, POLE and GLOBE.

4 Make a coloured layer for each part, my colours being: BASE: red, SUPPORT: green, POLE: magenta, GLOBE: blue.

5 The various parts of the model can be created by different 3D surface techniques, e.g.
    BASE: *P 94 a)* ruled surface circles
                 P105 b) revolve surface a polyline outline
    SUPPORT: *P94 a)* ruled surface
                      b) two ends with 3D face
    POLE: *P94 a)* ruled between two circles
                 P182 b) tabulated with a circle and line
                 P105 c) revolved with a line and circle
    GLOBE: *P105* revolve a semi-circle about a line.

6 The value of the SURFTAB system variable is important for the final appearance of the model, and I used:
    BASE: SURFTAB1 24
    SUPPORT: SURFTAB1 48
    POLE: SURFTAB1 6
    GLOBE: SURFTAB1 and SURFTAB2 both 60.

7 You will need to use several draw and modify commands before the surface can be added, e.g. copy, trim, etc.

8 One major hint – work with **UCS FRONT**.

9 My final layout has been zoomed centre with UCS BASE about the point 0,0,160 at 400 magnification.

10 If you manage to complete the 3D surface model, save it as R14MOD\GLOBE.

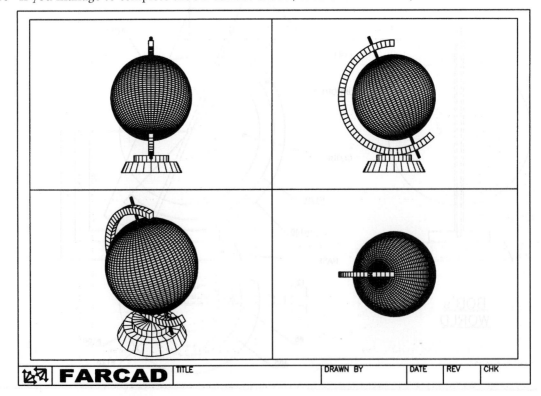

Figure 24.2   3D surface exercise – Bob's World.

Chapter 25

Dynamic viewing

Dynamic viewing is a powerful (yet underused) command which is very useful with 3D modelling as it allows models to be viewed from a perspective viewpoint. The command also allows objects to be 'cut-away' enabling the user to 'see inside' models. Dynamic viewing has its own terminology which is obvious when you are familiar with the command, but can be confusing to new users.

The basic concept of dynamic viewing is that the user has a **CAMERA** which is positioned at a certain **DISTANCE** from the model – called the **TARGET**. The user is looking through the camera lens at the model and can **ZOOM** in/out as required. The viewing direction is from the camera lens to a **TARGET POINT** on the model. The camera can be moved relative to the stationary target, and both the camera and target can be turned relative to each other. The target can also be **TWISTED** relative to the camera. Two other concepts which the user will encounter with the dynamic view command are the **slider bar** and the **perspective icon**. The slider bar allows the user to 'scale' the variable which is current, while the perspective icon is displayed when the perspective view is 'on'.

Figure 25.1(A) displays the various dynamic view concepts:
a) the basic terminology
b) the slider bar
c) the perspective icon.

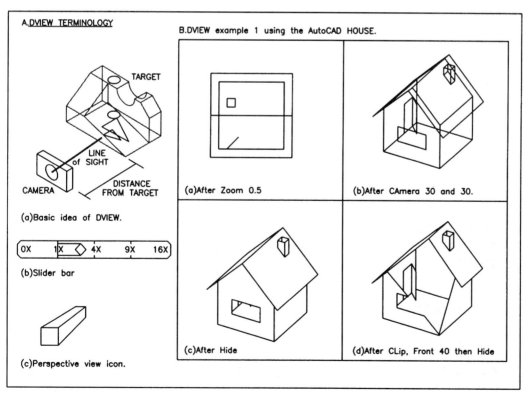

Figure 25.1 Dynamic view terminology and AutoCAD's 'House block'.

The dynamic view command has 12 options, these being:

CAmera,TArget,Distance,POints,Pan,Zoom,TWist,CLip,Hide,Off,Undo,eXit.

The option required is activated by entering the CAPITAL letters at the command line, e.g. CA for the camera option, TW for twist, etc. The various options will be investigated with two examples:

a) AutoCAD's dynamic view 'house'

b) a previously saved drawing.

Note: 1. Dynamic view is a model space concept, and cannot be used in paper space.
2. Dynamic view is **viewport independent**, i.e. if the command is used in a specific viewport, the model display in the other viewports will not be affected.

Example 1 – AutoCAD's 'house'

AutoCAD has a 'drawing' – actually a type of block – which can be used as an interactive aid with the dynamic view command. We will use this house block to demonstrate the various options so:

1 Begin a new 2D drawing and refer to Fig. 25.1(B).

2 From the menu bar select **View–3D Dynamic View** and:

prompt	Select objects
respond	**right-click** as nothing on the screen to select
a) *prompt*	Camera/Target/...
and	some cyan, red and black lines appear on the screen
enter	**Z <R>** – the zoom option
prompt	slider bar with scale
and	Adjust zoom scale factor<1>
enter	**0.5 <R>**
and	full plan view of house – fig. (a)
b) *prompt*	Camera/Target/...
enter	**CA <R>** – the camera option
prompt	*ghost image of house* which moves as mouse moved
and	Toggle angle in/Enter angle from XY plane
enter	**30 <R>**
prompt	Toggle angle from/Enter angle in XY plane from X axis
enter	**30 <R>**
and	3D view of house – fig. (b)
c) *prompt*	Camera/Target/...
enter	**H <R>** – the hide option
and	house displayed with hidden line removal – fig. (c)
d) *prompt*	Camera/Target/...
enter	**CL <R>** – the clip option
prompt	Back/Front/<Off>
enter	**F <R>** – the front clip option
prompt	Eye/ON/OFF/<Distance from target><1>
enter	**40 <R>**
prompt	Camera/Target/...
enter	**H <R>** – the hide option
and	house displayed 'cut-away' at front as fig. (d)
e) *enter*	**U <R>** – undoes the hide effect of (d)
enter	**U <R>** – undoes the clip effect of (d)
enter	**U <R>** – undoes the hide effect of (c)
and	**leave house with Camera option displayed**
and	**command prompt line options**
then	**read the explanation before proceeding**.

Explanation of the command

Dynamic view is an **interactive** command and the various options can be used one after the other. The undo (U) option will undo the last option performed, and can be used repeatedly until all the options entered have been 'undone'. Some of the options have been used to demonstrate how the command is used, these options being zoom, camera, clip, hide and undo. The hide option is very useful as it allows the model to be displayed when other options have been entered, and removes the 'ambiguity' effect from the model. The command can be used with all 3D models, i.e. extruded, wire-frame, surface and solid. The command is also **viewport independent**, i.e. it can be used in any viewport without affecting the display in other viewports. The AutoCAD 'house' is a user-reference, and if a model is displayed on the screen, this model will assume the house orientation when the dynamic view command is completed. This will be investigated in one of later examples.

The house displayed on the screen has been left with the Camera option with entered angles of 30 and 30. We will continue with the screen display and investigate the other dynamic view options. This means that you have to enter the various options and values as prompted.

CAmera

1 This option is used to direct the camera at the target and the camera can be 'tilted' relative to two planes with two angles:
   prompt 1: angle in the XY plane, between $-90°$ and $+90°$
   prompt 2: angle from the XY plane, between $-180°$ and $+180°$.

2 The angles can be:
   a) toggled using the ghost image as a guide
   b) entered directly from the keyboard.

3 Using the CAmera option enter the following angle values, the result being displayed in Fig. 25.2(A) with the HIDE effect:

angle in XY plane	*angle from XY plane*
a) 35	35
b) 35	-35
c) -35	35
d) -35	-35

4 The option can be considered similar to the **VPOINT ROTATE** command.

   When all the above entries have been completed, return the camera angles to the original 30 and 30, but do not exit the command.

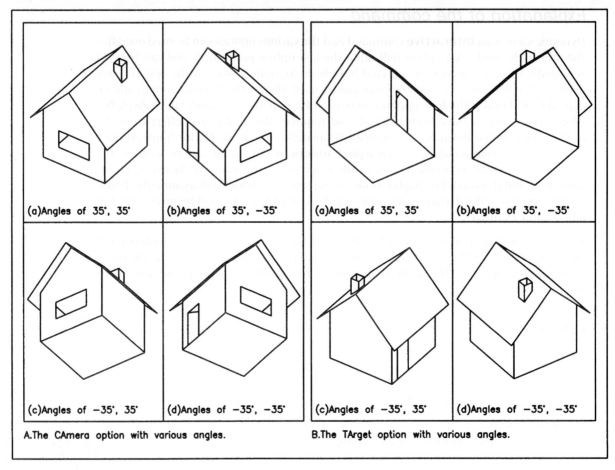

(a)Angles of 35°, 35°   (b)Angles of 35°, −35°   (a)Angles of 35°, 35°   (b)Angles of 35°, −35°

(c)Angles of −35°, 35°   (d)Angles of −35°, −35°   (c)Angles of −35°, 35°   (d)Angles of −35°, −35°

A.The CAmera option with various angles.   B.The TArget option with various angles.

Figure 25.2 The dynamic view CAmera and TArget options.

TArget

1 This option allows the target (the model) to be tilted relative to the camera. The two angle prompts are the same as the camera option:
prompt 1: angle in the *XY* plane
prompt 2: angle from the *XY* plane.

2 The angles can be toggled or entered.

3 Using the TArget option enter the following angle values, the result being displayed in Fig. 25.2(B) with hide:

	angle in XY plane	angle from XY plane
a)	35	35
b)	35	−35
c)	−35	35
d)	−35	−35

4 The option can be used to give the same effect as the camera option, but it should be remembered that the camera and target are being 'tilted' in the 'opposite sense' to each other.

5 When all angles have been entered, restore the camera to angles of 30 and 30, but do not exit the command.

TWist

1. A very useful option as it allows the 'plane' on which the target is 'resting' to be twisted through an entered angle. This angle can be positive or negative and have values between 0 and 360°.

2. The prompt with this option is **New view twist**.

3. The result of the option is dependent on the CAmera/TArget angles.

4. Using the TWist option enter angle values of:
   a) 35   *b*) −35   *c*) 180   *d*) −90.

5. Figure 25.3(A) displays the result of the TWist entries.

6. This is one of the few AutoCAD commands which allows models to be 'flipped' over by 180°.

7. When the four twist angles have been viewed with the hide effect, restore the original twist angle of 0, with the camera options of 30 and 30. Do not exit the command.

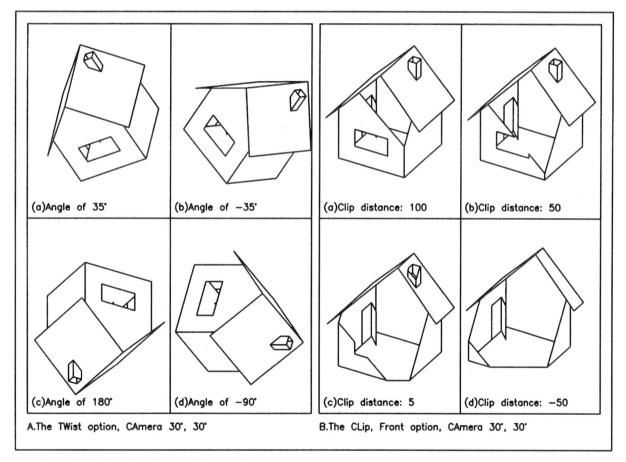

Figure 25.3 The dynamic view TWist and CLip options.

CLip

1 The clip option of the dynamic view command is probably of the most useful of all the options, as it allows models to be 'cut-away', thus allowing the user to 'see inside' the model.

2 The user selects a (F)ront or (B)ack clip and then decides on the clip distance either:
   a) using the slider bar
   b) entering a value at the command line.

3 The result of the clip option is dependent on the CAmera/TArget angles.

4 With the CAmera angles set to 30 and 30, enter the following *Front Clip* distances from the target:
   a) 100   *b*) 50   *c*) 5   *d*) –50.

5 The result of these clip entries is shown in Fig. 25.3(B).

6 When the clip option attempts are complete, undo the various effects with **U <R>** until the house is again displayed with the CAmera settings of 30 and 30.

POints

1 This option allows the model (the target) to be viewed from a specific 'standpoint': the user looking at a specific point on the target.

2 Two sets of coordinates are required:
   a) the target point coordinates to be looked at
   b) the coordinates of the camera – the user.

   The coordinate entries can be absolute or relative.

4 When this option is used, the PAn option is sometimes also required.

5 The result does not depend on the CAmera or TArget options.

6 Figure 25.4 displays the eight different POints entries, these being:

Target point	*Camera point*
a) 0,0,0	1,0,0
b) 0,0,0	0,1,0
c) 0,0,0	0,0,1
d) 0,0,0	1,1,0
e) 0,0,0	1,0,1
f) 0,0,0	1,1,1
g) 1,2, 3	0,0,0
h) 0,0,0	1,2,3

7 The option is similar to the **VPOINT VECTOR** command.

8 When all the points entries have been entered, restore the camera angles of 30 and 30, but do not exit the command.

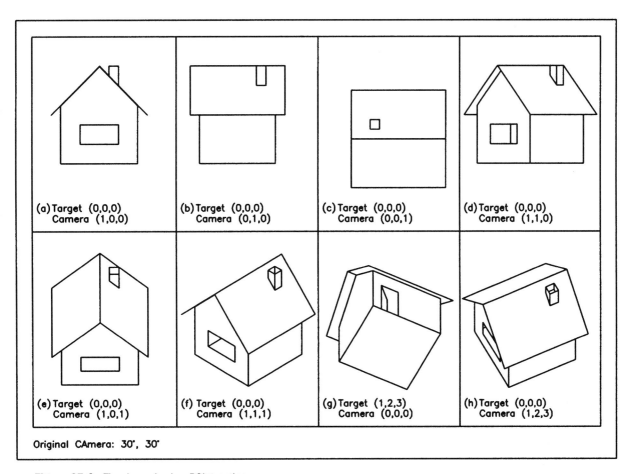

Figure 25.4 The dynamic view POint option.

Distance

1 Alters the distance between the camera and the target.

2 The distance can be:
a) entered as a value from the command line
b) toggled using the slider bar.

3 Figure 25.5(A) displays four distance entries with the camera set to 30 and 30, the entered distances being:
a) 1000   *b*) 1500   *c*) 2500   *d*) 5000.

4 The distance option introduces **true perspective** to the model.

5 When the distance option has been used, and the command is exited, the zoom command cannot be used.

Zoom

1 This option does what you would expect – it 'zooms the model'.

2 The zoom factor can be:
   a) entered as a value from the keyboard
   b) toggled using the slider bar.

3 Figure 25.5(B) displays the house at camera angles of 30 and 30 with zoom factors of:
   a) 1   *b*) 0.75   *c*) 0.5   *d*) 0.25.

Pan

1 This option is similar to the AutoCAD PAN command, but the 'real-time' pan effect is not available.

2 The user selects (or enters) the pan displacement.

Hide

1 Will display the model with a hide effect.

2 Removes any ambiguity.

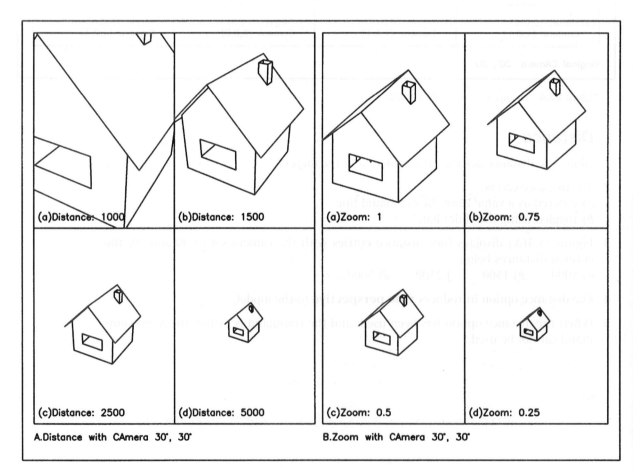

Figure 25.5 The dynamic view Distance and Zoom options.

Undo

1 Entering **U <R>** will undo the last option.

2 Can be used repetitively until all the option entries have been undone.

eXit

1 The **X <R>** option will end the dynamic view command and a blank screen will be returned.

2 The blank screen results because we did not have any drawing displayed, the AutoCAD 'house' being a visual aid indicating what any model would 'look like'.

3 With 'real models', the model orientation will be similar to the 'house' orientation, as will now be investigated.

Example 2

In this example we will use the dynamic view command with a previously created model.

1 Open the ruled surface model R14MOD\ARCHES created during Chapter 16 and refer to Fig. 25.6.

2 Make the upper left viewport active.

3 Menu bar with **View–3D Dynamic View** and:
   prompt   Select objects and right-click
   prompt   AutoCAD's house as an 'end view' in the active viewport
   and   dynamic view options
   enter   **CA <R>**
   prompt   Angle from XY plane and enter: **35 <R>**
   prompt   Angle in XY plane and enter: **25 <R>**
   prompt   dynamic view options and enter: **X <R>**.

4 The arch model will be displayed with house CAmera configuration as fig. (b).

5 With the top right viewport active, enter **DVIEW <R>** at the command line and:
   prompt   Select objects
   respond **window the model then right-click**
   prompt   dynamic view options
   enter   **TW <R>**
   prompt   New view twist and enter: **–90 <R>**
   prompt   dynamic view options and enter: **X <R>**
   and   Model displayed with new twist as fig. (c).

6 Use the dynamic view command in the lower viewports with the following entries:
Lower left	*Lower right*
options: TArget	options: CAmera
angles: –40 and 30	angles: 30 and 30
options: CLip, Front	options: TWist, angle: 180
distance: 10	options: CLip, Front, Distance: 30
fig. (d)	fig. (e).

7 Save the new viewport configuration, but we will not refer to this exercise again.

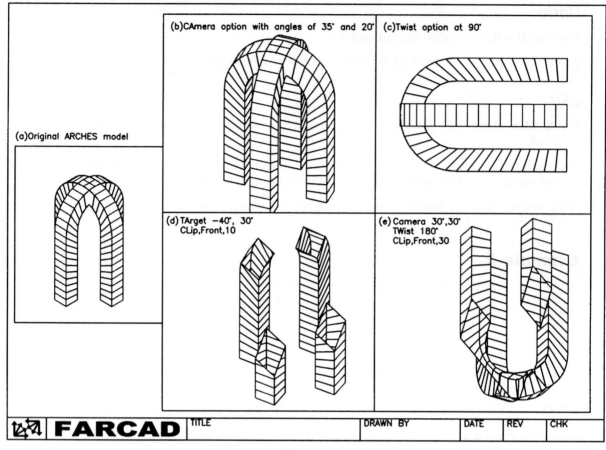

Figure 25.6 Dynamic view example 2 using R14MOD\ARCHES.

Summary

1 The dynamic view command is viewport specific, i.e. it only affects the active viewport.

2 The command has several useful options:
CAmera, TArget: similar to VPOINT rotate
TWist: allows models to be 'inverted'
Distance: introduces true perspective
CLip: useful to 'see inside' models.

3 The command can be activated from:
a) menu bar with View–3D Dynamic View
b) keyboard entry with DVIEW.

4 The command can be used:
a) directly on models
b) interactively using AutoCAD's house.

5 The command is used relative to the WCS – observe the prompt line when the command is activated.

Assignment

No specific activity, but investigate the command with some previously created models.

Chapter 26

Viewport-specific layers

When layers are used with multiple viewports they are generally **global**, i.e. what is drawn on a layer in one viewport will be displayed in the other viewports. This is quite acceptable for creating models but is unacceptable for certain other concepts, e.g. when dimensions have to be added to a model, or if the model is to be sectioned. If dimensions are to be added to a multi-view drawing, then these dimensions should only be visible in the active viewport.

In this chapter we will investigate how to create and use viewport-specific layers.

Viewport-specific layer example
Global layers

1 Open the 3D-faced model R14MOD\CHEESE from Chapter 14 and refer to Fig. 26.1.

2 With UCS BASE, lower right viewport active make layer DIM current and display the Dimension toolbar.

3 Select the LINEAR DIMENSION icon from the Dimension toolbar and dimension lines 1–2 and 1–3.

4 Make the upper right viewport active and restore UCS FRONT.

5 Linear dimension line a–b and baseline dimension lines a–c and a–d.

6 The five dimensions will be displayed in all four viewports due to the GLOBAL nature of layer DIM. Figure 26.1(a) displays the five dimensions as added to the 3D viewport.

7 Now erase the five dimensions and restore UCS BASE.

8 Remember that dimensioning is a 2D concept, the result depending on the orientation of the UCS.

Viewport specific layers

1 Still with the 3D faced model displayed on screen?

2 Menu bar with **Format–Layer** and:

  prompt   Layer & Linetype Properties dialogue box
  respond   1. select the Dim layer line – turns blue
              2. pick New three times to make three new magenta layers, Layer1, Layer2 and Layer3
              3. Alter the new layer names to:
                  Layer1: DIMTL   Layer2: DIMTR   Layer3: DIMBR
              4. The nomenclature for these new layer names is for individual viewports, e.g. TL: top left, BR: bottom right, TR: top right
              5. pick OK.

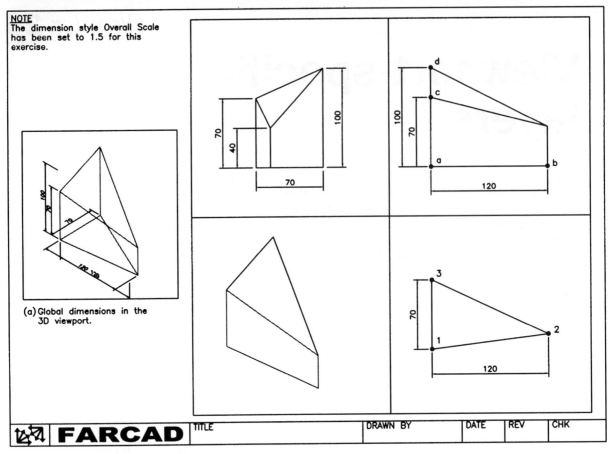

NOTE
The dimension style Overall Scale has been set to 1.5 for this exercise.

(a) Global dimensions in the 3D viewport.

FARCAD

Figure 26.1 Viewport specific layer example using R14MOD\CHEESE.

3 In model space make the top left viewport active and:
   a) activate the Layer Properties dialogue box
   b) by selecting the Freeze/Thaw in current viewport icon, freeze the new layers Dimtr and Dimbr
   c) pick OK.

4 With the top right viewport active:
   a) activate the Layer Properties dialogue box
   b) pick the currently freeze icon and freeze layers Dimtl and Dimbr
   c) pick OK.

5 With the lower right viewport active use the Layer Properties dialogue box to currently freeze layers Dimtl and Dimtr.

6 In the lower left viewport currently freeze the three new layers, i.e. Dimtl, Dimtr and Dimbr.

7 With UCS BASE and the lower right viewport active:
   a) make layer DIMBR current
   b) linear dimension lines 1–2 and 1–3
   c) the two dimensions should only be displayed in the lower right viewport?

8 Restore UCS FRONT and make the upper right viewport active and:
   a) make layer DIMTR current
   b) linear dimension line a–b and baseline dimension lines a–c and a–d
   c) the three dimensions are only displayed in the top right viewport.

9 *a*) Restore UCS RIGHT
  b) make the upper left viewport active
  c) make layer DIMTL current
  d) add the four dimensions as Fig. 26.1.

10 *Notes*.
  a) All dimensions should have been added to the viewport which was active.
  b) The dimensions have been added to a viewport specific layer which was made current in the viewport where the dimensions had to be added.
  c) Generally the 3D viewport does not require a dimension layer as dimensions are not usually added to a 3D viewport.
  d) In the 3D viewport the three viewport specific layers were all currently frozen.
  e) I set the Overall Dimension scale (Dimension Style–Geometry) to 1.5 in this exercise.

Layer states

1 Layers can have different **states** depending on whether they are global or viewport specific, these states being easily controlled:
  a) from the Layer and Linetype Properties dialogue box
  b) from the icons displayed in the Object Properties toolbar.

2 The different states are:

Global	Viewport specific
On/Off	On/Off
Freeze/Thaw	Freeze/Thaw in all viewports
Lock/Unlock	Freeze/Thaw in current viewports
	Freeze/Thaw in new viewports
	Lock/Unlock

3 Layers can have more than one state active at a time, e.g. on and locked.

4 Figure 26.2 displays the Layer Properties dialogue box layers icons and names.

5 *a*) An icon displayed in yellow is on or thawed
  b) an icon displayed in blue is off or frozen.

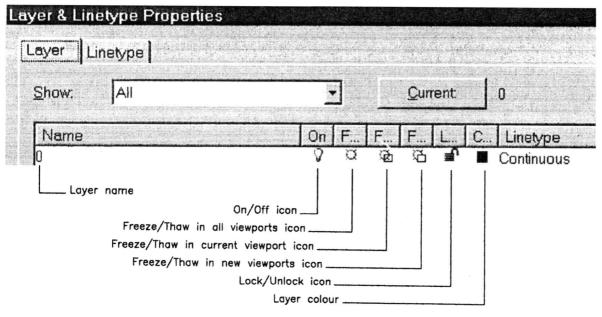

Figure 26.2 The Layer Properties dialogue box layer state icons.

Summary

1 Viewport specific layers are layers which are specific to a named viewport.

2 Viewport specific layers are used with multi-view drawings and are essential for such concepts as dimensioning a 3D model.

3 Viewport specific layers can **only be created** if the TILEMODE variable is set to 0, i.e. if paper space is active.

4 Viewport specific layers can be created using the:
   a) Layer Properties dialogue box – recommended
   b) VPLAYER command – not considered in this exercise.

5 The most commonly used state with viewports specific layers is 'Freeze in current viewport'.

Assignment
Activity 17: Dimensioning a 3D model

1 Recall the 3D-faced hexagonal prism from Activity 10.

2 Make three new dimension layers as the exercise, e.g. DIMTL, etc.

3 Currently freeze 'unwanted layers' in the appropriate viewports using the same method as the example.

4 Making the appropriate viewport active:
   a) restore the required UCS
   b) make the correct dim layer current
   c) add the dimensions given.

5 Save the completed activity.

Chapter 27

Introduction to solid modelling

Three-dimensional modelling with computer-aided draughting and design (CADD) can be considered as three categories:

a) wire-frame modelling
b) surface modelling
c) solid modelling.

We have already created wire-frame and surface models and will now concentrated on how solid models are created.

This chapter will summarize the three model types.

Wire-frame modelling

1 Wire-frame models are defined by points and lines and are the simplest possible representation of a 3D component. They may be adequate for certain 3D model representation and require less memory than the other two 3D model types, but wire-frame models have several limitations:

a) *Ambiguity*: it is difficult to know how a wire-frame model is being viewed, i.e. from above or from below?
b) *No curved surfaces*: while curves can be added to a wire-frame model as arcs or trimmed circles, an actual curved surface cannot. Lines may be added to give a 'curved effect' but the computer does not recognize these as being part of the model.
c) *No interference*: as wire-frame models have no surfaces, they cannot detect interference between adjacent components. This makes them unsuitable for kinematic displays, simulations, etc.
d) *No physical properties*: mass, volume, centre of gravity, moments of inertia, etc. cannot be calculated.
e) *No shading*: as there are no surfaces, a wire-frame model cannot be shaded or rendered.
f) *No hidden line removal*: as there are no surfaces, it is not possible to display the model with hidden line removal.

2 AutoCAD Release 14 allows wire-frame models to be created.

Surface modelling

1 A surface model is defined by points, lines and faces. A wire-frame model can be 'converted' into a surface model by adding these 'faces'. Surface models have several advantages when compared to wire-frame models, some of these being:

a) Recognition and display of curved profiles.
b) Shading, rendering and hidden line removal are all possible, i.e. no ambiguity.
c) Recognition of holes.

2 Surface models are suited to many applications but they have some limitations which include:
   a) *No physical properties*: other than surface area, a surface model does not allow the calculation of mass, volume, centre of gravity, moments of inertia, etc.
   b) *No detail*: a surface model does not allow section detail to be obtained.

3 Several types of surface model can be generated including:
   a) plane and curved swept surfaces
   b) swept area surfaces
   c) rotated or revolved surfaces
   d) splined curve surfaces
   e) nets or meshes.

4 AutoCAD Release 14 allows surface models of all types to be created.

Solid modelling

1 A solid model is defined by the volume the component occupies and is thus a real 3D representation of the component. Solid modelling has many advantages which include:

   a) Complete physical properties of mass, volume, centre of gravity, moments of inertia, etc.
   b) Dynamic properties of momentum, angular momentum, radius of gyration, etc.
   c) Material properties of stress–strain.
   d) Full shading, rendering and hidden detail removal.
   e) Section views and true shape extraction.
   f) Interference between adjacent components can be highlighted.
   g) Simulation for kinematics, robotics, etc.

2 Solid models are created using a **solid modeller** and there are several types of solid modeller, the two most common being:
   a) Constructive solid geometry or constructive representation, i.e. CSG/CREP. The model is created from solid primitives and/or swept surfaces using Boolean operations.
   b) Boundary representation (BREP). The model is represented by the edges and faces making up the surface, i.e. the topology of the component.
   c) AutoCAD Release 14 supports solid models of the CSG/CREP type.
   d) The AutoCAD Release 14 modeller is based on the **ACIS** solid modeller which has certain advantages over the R12 AME modeller.

Comparison of the model types

The three model types are displayed in:
a) Figure 27.1: as models with hidden line removal
b) Figure 27.2: as model cross-sections.

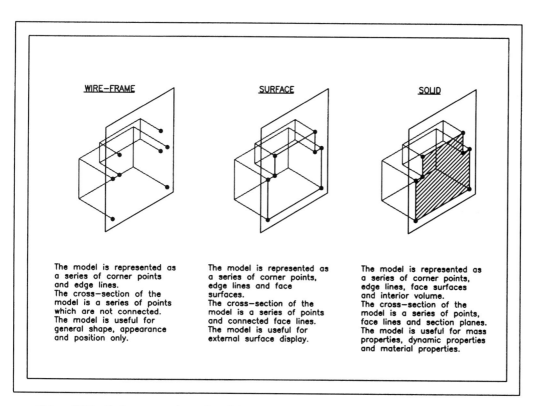

3D WIRE-FRAME MODEL

1. Model has length, width and height.
2. There are no surfaces on the model.
3. The model does not have area, mass or volume.
4. HIDE has no effect on the appearance of the model.
5. The model displays AMBIGUITY ie it is difficult to know if it is being viewed from above or from below.

3D SURFACE MODEL

1. Model has length, width and height.
2. Surfaces can be added.
3. The model has a surface area but no mass or volume.
4. HIDE can be used.
5. There is no ambiguity.
6. No sections can be extracted from the model.

3D SOLID MODEL

1. Model has length, width and height.
2. Model has a surface area, mass and volume.
3. HIDE can be used.
4. There is no ambiguity.
5. Model has mass properties eg centroid, moment of inertia, radius of gyration etc.
6. The model is 'real' and sections can be extracted from it.

Figure 27.1   Simple comparison between wire-frame, surface and solid models.

WIRE-FRAME

The model is represented as a series of corner points and edge lines.
The cross-section of the model is a series of points which are not connected.
The model is useful for general shape, appearance and position only.

SURFACE

The model is represented as a series of corner points, edge lines and face surfaces.
The cross-section of the model is a series of points and connected face lines.
The model is useful for external surface display.

SOLID

The model is represented as a series of corner points, edge lines, face surfaces and interior volume.
The cross-section of the model is a series of points, face lines and section planes.
The model is useful for mass properties, dynamic properties and material properties.

Figure 27.2   Further comparison of model types as cross-section.

The solid model standard sheet

A solid model standard sheet (prototype drawing) will be created as a template file using the layout from the surface model exercises, i.e. MV3DSTD. This standard sheet will:
a) be for A3 paper
b) have a four viewport (paper space) configuration.

1 Menu bar with **File–New** and open your MV3DSTD template file.

2 Check the following:
   a) Tool–UCS–Named UCS to display BASE, FRONT, RIGHT
   b) Layers MODEL (current), OBJECTS, SECT, TEXT, VP, SHEET, DIM
   c) Sheet layout to your own specification
   d) Text style: ST1 with romans.shx font
   e) Dimension style 3DSTD with various settings.

3 At the command line enter **PURGE <R>** and:
prompt	`Purge unwanted Blocks/...`
enter	**LA <R>** – layer option
prompt	`Names to purge<*>` and right-click
prompt	`Verify each name to be purges<Y>` and right-click
prompt	`Purge layer DIM<N>` and right click
prompt	`Purge layer OBJECTS` and enter: **Y <R>**
prompt	`Purge layer SECT<N>` and right-click
prompt	`Purge layer TEXT<N>` and right-click.

4 At the command line enter **ISOLINES <R>** and:
prompt	`New value for ISOLINES<4>`
enter	**24 <R>**.

5 At the command line enter **FACETRES <R>** and:
prompt	`New value for FACETRES<0.5000>`
enter	**1 <R>**.

6 Display the Draw, Modify, Object Snap, Dimension and Solids toolbars.

7 Make the lower left viewport active.

8 Menu bar with **File–Save As** and:
prompt	Save Drawing As dialogue box
respond	1. scroll at Save as Type
	2. pick **Drawing Template File (*.dwt)**
	3. enter file name as: **A3SOL.dwt**
	4. pick **Save**
prompt	`Template Description` dialogue box
enter	**My solid model prototype layout created on ???**
then	pick OK.

9 You are now ready to start creating solid models.

Notes

1   Two new system variables have been introduced in the creation of the A3SOL template file, these being:

ISOLINES: controls the number of tessellation lines used in the visualization of curved portions of models. It is an integer with values between 0 and 2047. The default value is 4.

FACETRES: adjusts the smoothness of shaded and hidden line removal models. The value can be between 0.01 and 10.00, the default being 0.5.

2   The ISOLINES system variable is similar to the SOLWDENS variable in the R12 AME solid modeller.

3   The ISOLINES and FACETRES values may be altered with some of our created models.

4   The viewports have not been zoom-centred, as this will depend on the model being created.

5   Solid modelling consists of creating 'composites' from 'primitives' and Release 14 has the following types of primitive:
a) basic
b) swept
c) edge.

6   All three types of primitive will be investigated with examples and the various options for each will be discussed.

Chapter 28

The basic solid primitives

AutoCAD Release 14 supports the six basic solid primitives of box, wedge, cylinder, cone, sphere and torus. In this chapter we will create (interesting?) layouts using each primitive and also investigate the various options which are available.

During the exercises do not just accept the coordinate values given, try and reason out why they are being used.

The BOX primitive – Fig. 28.1

1 Open your A3SOL template file with layer MODEL, UCS BASE and the lower left viewport active. Solids toolbar displayed?

2 Select the BOX icon from the Solids toolbar and:

   prompt   Center/<Corner of box><0,0,0>
   enter   **0,0,0 <R>**
   prompt   Cube/Length/<other corner>
   enter   **C <R>** – the cube option
   prompt   Length
   enter   **100 <R>**
   and   a red cube is displayed in all viewports.

3 Select from the menu bar **Draw–Solids–Box** and:

   prompt   Center/<Corner of box><0,0,0>
   enter   **100,0,0 <R>**
   prompt   Cube/Length/<other corner>
   enter   **L <R>** – the length option
   prompt   Length and enter: **40 <R>**
   prompt   Width and enter: **80 <R>**
   prompt   Height and enter: **30 <R>**
   and   another red cuboid is displayed in all viewports.

4 Use the Properties icon from the Object Properties toolbar to change the colour of this cuboid to green.

5 Now zoom centre about 60,20,60 at 200 magnification in all viewports.

6 At the command line enter **BOX <R>** and:

   prompt   Center/<Corner of box><0,0,0>
   respond   **Endpoint icon and pick pt1**
   prompt   Cube/Length/<other corner>
   enter   **@120,60,15 <R>** – the diagonally opposite corner point.

7 Change the colour of this box to blue using either the Properties icon or CHNAGE at the command line.

8 Restore UCS RIGHT.

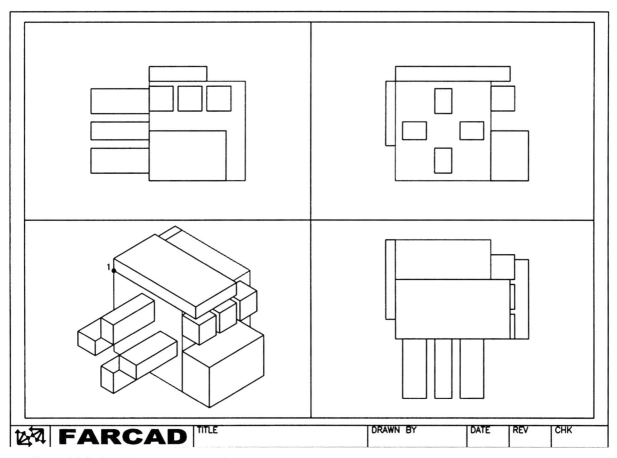

FARCAD | TITLE | DRAWN BY | DATE | REV | CHK

Figure 28.1 The BOX solid primitive (BOXPRIM).

9 Create a solid box with the following:
 a) corner: 0,70,100
 b) cube option with length: 25
 c) colour: magenta.

10 Rectangular array the magenta box:
 a) for 1 row and 3 columns
 b) with column distance: 30.

11 Restore UCS FRONT.

12 Activate the solid BOX command and:
 prompt   Center/<Corner of box><0,0,0>
 enter   **C <R>** – the box center point option
 prompt   Center of box and enter: **50,80,30 <R>**
 prompt   Cube/Length/<other corner> and enter: **L <R>**
 prompt   Length and enter: **18 <R>**
 prompt   Width and enter: **25 <R>**
 prompt   Height and enter: **60 <R>**.

13 Change the colour of this last box to suit yourself (I used colour number 20) then polar
 array it with:
 a) center point: 50,50
 b) items: 4
 c) 360 angle with rotation.

14 Finally restore UCS BASE and create another solid box with:
a) corner: 0,100,100
b) length: –10   width: –80   height: –65
c) colour: cyan.

15 The model is now complete, so hide each viewport with menu bar and **View–Hide** or enter **HIDE <R>** at the command line.

16 Shade each viewport with **SHADE <R>** at the command line or with the menu bar sequence **View–Shade**. The model will be displayed as coloured blocks – impressive?

17 Save the layout as R14MOD\BOXPRIM.

18 *Investigate*.
a) the four SHADE options from the View menu bar
b) the REGEN and REGENALL commands after HIDE (View menu bar).

The WEDGE primitive – Fig. 28.2

1 Open the A3SOL template file with layer MODEL, UCS BASE and lower left viewport active.

2 Menu bar with **Draw–Solids–Wedge** and:
prompt	Center/<Corner of wedge><0,0,0>
enter	**0,0,0 <R>**
prompt	Cube/Length/<other corner>
enter	**C <R>** – the cube option
prompt	Length
enter	**100 <R>**
and	red wedge displayed in all viewports.

3 Select the WEDGE icon from the Solids toolbar and:

prompt	Center/<Corner of wedge><0,0,0>
enter	**0,0,0 <R>**
prompt	Cube/Length/<other corner>
enter	**@80,–60 <R>** – the other corner option
prompt	Height and enter: **50 <R>**.

4 Change the colour of this wedge to blue.

5 In each viewport, zoom centre about 80,30,50 at 200 magnification and pan to suit if needed.

6 At the command line enter **WEDGE <R>** and create a solid wedge with:
a) corner: 100,100,0
b) length: 60   width: 60   height: 80
c) colour: green
d) 2D rotate the green wedge about the point 100,100 by +**90°**.

7 Restore UCS FRONT and create a wedge:
a) corner: **endpoint icon and pick pt1**
b) length: –50
c) width: –100
d) height: 30
e) colour: magenta.

8 Restore UCS BASE.

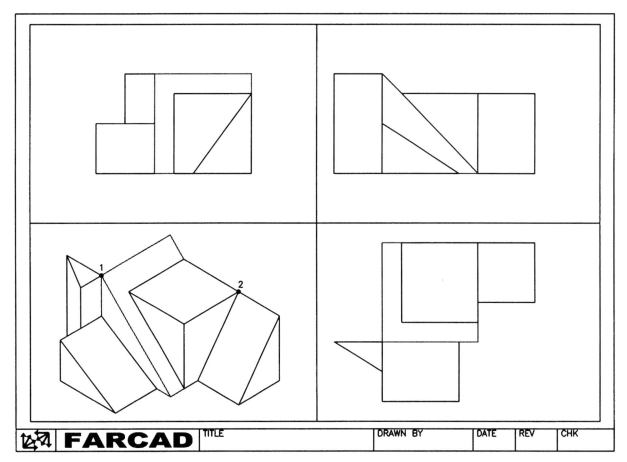

Figure 28.2 The WEDGE solid primitive (WEDPRIM).

9 The final wedge is to be created with:
   a) corner: **endpoint of pt2**
   b) cube option with length: −80
   c) colour: cyan.

10 Hide and shade in each viewport.

11 Regenall and save as R14MOD\WEDPRIM.

The CYLINDER primitive – Fig. 28.3

1 Open A3SOL, UCS BASE, layer MODEL, lower left active.

2 Menu bar with **Draw–Solids–Cylinder** and:

    prompt   Elliptical/<center point><0,0,0>
    enter   **0,0,0 <R>** – the cylinder base centre point
    prompt   Diameter/<Radius>
    enter   **60 <R>** – the cylinder radius
    prompt   Center of other end/<Height>
    enter   **100 <R>** – the cylinder height
    and   a red cylinder is displayed, centred on 0,0,0.

3 Select the CYLINDER icon from the Solids toolbar and:

    prompt   Elliptical/<center point><0,0,0>
    enter   **E <R>** – the elliptical option
    prompt   Center/<Axis endpoint>
    enter   **C <R>** – the centre option
    prompt   Center of ellipse<0,0,0>
    enter   **80,0,0 <R>**
    prompt   Axis endpoint and enter: **@20,0,0 <R>**
    prompt   Other axis distance and enter: **@0,30,0 <R>**
    prompt   Center of other end/<Height>
    enter   **50 <R>**.

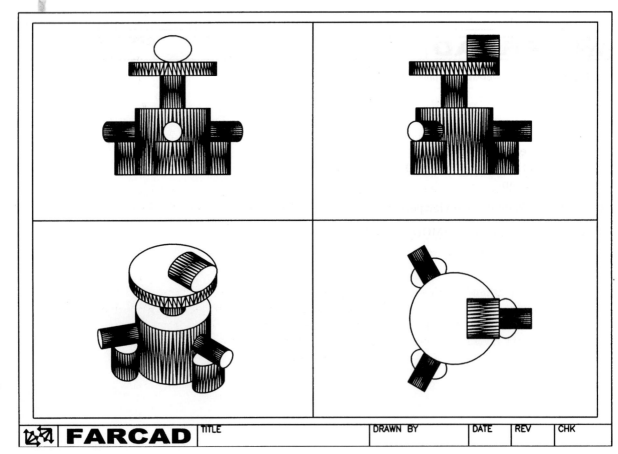

FARCAD | TITLE | DRAWN BY | DATE | REV | CHK

Figure 28.3 The CYLINDER solid primitive (CYLPRIM).

4 Change the colour of this cylinder to green, then polar array it:
  a) about the point 0,0
  b) for three items, full angle with rotation.

5 Zoom centre about 0,0,80 at 300 magnification in all viewports.

6 At the command line enter **CYLINDER <R>** and:
  prompt   Elliptical/<center point><0,0,0>
  enter   **60,0,65 <R>** – the centre point
  prompt   Diameter/<Radius> and enter: **15 <R>**
  prompt   Center of other end/<Height>
  enter   **C <R>** – centre of other end option
  prompt   Center of other end
  enter   **@60,0,0 <R>**.

7 Change the colour of this cylinder to magenta and polar array it with the same entries as step (4).

8 Create another two cylinders:

centre pt	rad	ht	colour
0,0,100	20	50	cyan
0,0,150	70	20	blue

9 Finally create an elliptical cylinder:
  a) centre: 70,0,190
  b) axis endpoint: @30,0
  c) other axis distance: @0,20
  d) centre of other end: @–50,0,0
  e) colour to suit.

10 Hide the model and note the triangular facets which are not displayed when the cylinder primitives are created. Shade then regen.

11 Save the model as R14MOD\CYLPRIM.

12 *Investigate*.
  a) With the 3D viewport active, enter ISOLINES at the command line and enter a value of 6. Hide the model and note the effect.
  b) change the ISOLINES value to 48 and hide.
  c) return the ISOLINES value to the original 24.
  d) enter FACETRES at the command line and alter the value to 2, then hide and shade – note effect.
  e) alter FACTRES to 5, hide and shade (takes longer).
  f) return FACETRES to 1.

13 *Note*.
  The ISOLINES system variable controls the appearance of primitive curved surfaces when they are created. The triangulation effect, or FACETS, is controlled by the system variable FACETRES. The higher the value of FACETRES (max 10) then the 'better the appearance' of curved surfaces, but the longer it takes for hide and shade. At our level, the values of 24 for ISOLINES and 1 for FACETRES are sufficient.

The CONE primitive – Fig. 28.4

1 Open the A3SOL template file, UCS BASE, etc.

2 Menu bar with **Draw–Solid–Cone** and:
   prompt　　`Elliptical/<center point><0,0,0>`
   enter　　　**0,0,0 <R>** – the cone base centre point
   prompt　　`Diameter/<Radius>` and enter: **50<R>**
   prompt　　`apex/<Height>` and enter: **60 <R>**.

3 Create another cone with:
   a) centre: 0,0,0
   b) radius: 90
   c) height: −80
   d) colour: green.

4 Select the CONE icon from the Solids toolbar and:
   prompt　　`Elliptical/<center point><0,0,0>`
   enter　　　**E <R>** – the elliptical option
   prompt　　`Center/<Axis endpoint>`
   enter　　　**C <R>** – the centre point option
   prompt　　`Center of ellipse<0,0,0>`
   enter　　　**70,0,0 <R>**
   prompt　　`Axis endpoint` and enter: **@20,0,0 <R>**
   prompt　　`Other axis distance` and enter: **@0,25,0 <R>**
   prompt　　`Apex/<Height>` and enter: **35 <R>**.

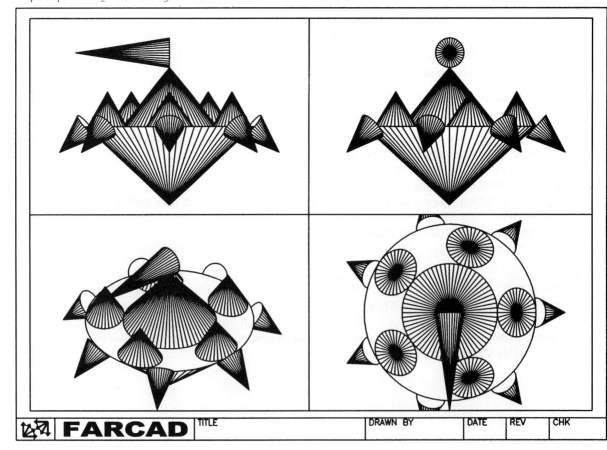

Figure 28.4 The CONE solid primitive (CONPRIM).

5 Change the colour of this cone to blue, then polar array it with:
   a) centre point: 0,0
   b) items: 5
   c) full angle to fill, with rotation.

6 Centre each viewport about 0,0,10 at 200 magnification.

7 At the command line enter **CONE <R>** and:
   prompt   Elliptical/<center point> and enter: **90,0,0 <R>**
   prompt   Diameter/<Radius> and enter: **15 <R>**
   prompt   Apex/<Height>
   enter   **A <R>** – the apex option
   prompt   Apex
   enter   **@40,0,0 <R>**.

8 Change the colour of this last cone to cyan.

9 Use the 3D ROTATE command with the cyan cone and:
   a) enter a *Y* axis rotation
   b) enter 90,0,0 as the point on the *Y* axis
   c) enter 41.63 as the rotation angle – why this figure?

10 Polar array the cyan cone about the point 0,0 for seven items, full angle with rotation.

11 Create the final cone with:
   a) centre: 0,0,75
   b) radius: 15
   c) apex option: @0,–100,0
   d) colour to suit.

12 Hide, shade then save as R14MOD\CONPRIM.

The SPHERE primitive – Fig. 28.5

1 Open the A3SOL template file, UCS BASE, etc.

2 Menu bar with **Draw–Solids–Sphere** and:
   prompt   Center of sphere<0,0,0>
   enter   **0,0,0 <R>**
   prompt   Diameter/<radius> of sphere
   enter   **60 <R>**.

3 Centre each viewport about 0,0,20 at 200 magnification.

4 Select the SPHERE icon from the Solids toolbar and:
   prompt   Center of sphere<0,0,0>
   enter   **80,0,0 <R>**
   prompt   Diameter/<Radius>
   enter   **D <R>** – the diameter option
   prompt   Diameter and enter: **40 <R>**.

5 Change the colour of this sphere to green, then polar array it:
   a) about the point: 0,0
   b) for five items
   c) full angle with rotation.

6 At the command line enter **SPHERE <R>** and create a sphere with:
   a) centre: 0,0,75
   b) radius: 15
   c) colour: cyan.

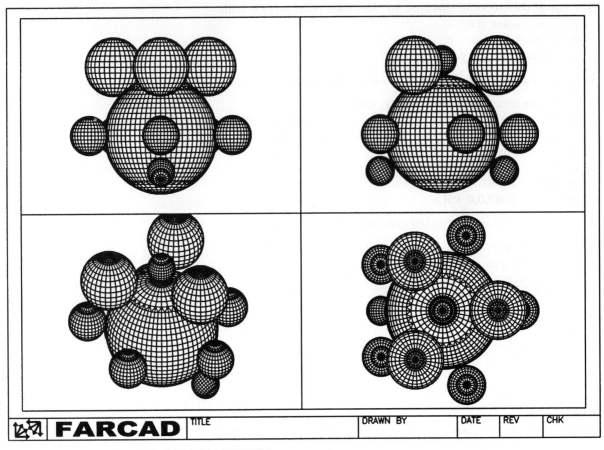

Figure 28.5 The SPHERE solid primitive (SPHPRIM).

7   Restore UCS FRONT and polar array the cyan sphere about the point 0,0 for three items with full angle rotation.

8   Restore UCS BASE and create the final sphere with:
    a) centre: 58,0,70
    b) radius: 30
    c) colour blue
    d) polar arrayed about 0,0 for three items, full angle rotation.

9   Hide, shade, save as R14MOD\SPHPRIM.

10 *Investigate*.
    a) ISOLINES set to 48 and hide, shade
    b) FACETRES set to 5 and hide, shade
    c) Note the appearance of the sphere primitives with these higher system variable values.

The TORUS primitive – Fig. 28.6

1 A3SOL template file with UCS BASE.

2 Menu bar with **Draw–Solids–Torus** and:
   prompt   Center of torus<0,0,0>
   enter   **0,0,0 <R>**
   prompt   Diameter/<Radius> of torus
   enter   **80 <R>**
   prompt   Diameter/<radius> of tube
   enter   **15 <R>**.

3 Restore UCS FRONT and select the TORUS icon from the Solids toolbar and:
   prompt   Center of torus<0,0,0>
   enter   **80,0,0 <R>**
   prompt   Diameter/<Radius> of torus and enter: **50 <R>**
   prompt   Diameter/<Radius> of tube and enter: **20 <R>**.

4 Change the colour of this torus to blue.

5 Restore UCS BASE and polar array the blue torus about 0,0 for three items with full circle rotation.

6 Restore UCS RIGHT and enter **TORUS <R>** at the command line and create a torus with:
   a) centre: 0,0,–95
   b) radius of torus: 80
   c) radius of tube: 20
   d) colour: green.

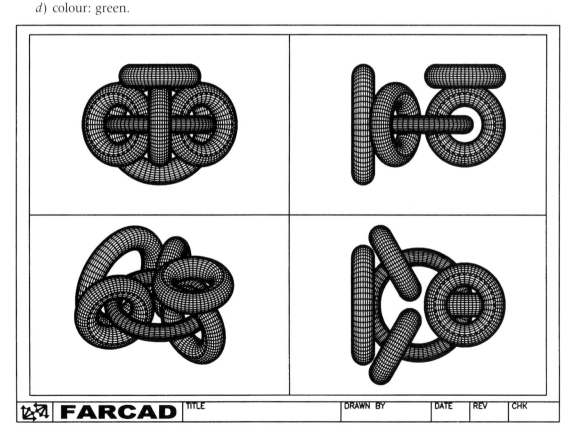

FARCAD | TITLE | | DRAWN BY | DATE | REV | CHK

Figure 28.6 The TORUS solid primitive (TORPRIM).

7 Restore UCS BASE and create the final torus with:
   a) centre: 80,0,80
   b) radius of torus: 50
   c) radius of tube: 20
   d) colour: to suit.

8 Finally, zoom centre about 0,0,0 at 300 magnification in all viewports.

9 Hide, shade and save as R14MOD\TORPRIM.

10 *Investigate*.
    The effect of ISOLINES and FACETRES on the layout.

Summary

1 The six solid primitives can be activated:
   a) from the menu bar with Draw–Solids
   b) by icon selection from the Solids toolbar
   c) by entering the solid name at the command line.

2 The corner/centre start points can be:
   a) entered as coordinates from the keyboard
   b) referenced to existing objects.

3 All six primitives have options:
   box: *a*) corner; centre
              b) cube; length, width, height; other corner
   wedge: *a*) corner; centre
              b) cube; length, width, height; other corner
   cylinder: *a*) circular; elliptical
              b) diameter; radius
              c) height; centre of other end
   cone: *a*) circular; elliptical
              b) diameter; radius
              c) height; centre of other end
   sphere: *a*) centre only
              b) diameter; radius
   torus: *a*) centre only
              b) diameter; radius of torus
              c) diameter; radius of tube.

Assignment
Activity 18: A solid primitive layout

Create a layout **of your own design** for the six basic primitives using the following information:

Box:	length: 100	*Wedge*:	length: 100	*Cylinder*:	radius: 40
	width: 100		width: 100		height: 80
	height: 80		height: 80		colour: green
	colour: red		colour: yellow		
Cone:	radius: 50	*Sphere*:	radius: 30	*Torus*:	torus rad: 80
	height: 100		colour: magenta		tube rad: 20
	colour: blue				colour: cyan

Chapter 29

The swept solid primitives

Solid models can be generated by extruding or revolving 'shapes' and in this chapter we will use several exercises to demonstrate how complex solids can be created from relatively simple shapes.

Extruded solids

Solid models can be created by extruding **closed objects** such as polyline shapes, polygons, circles, ellipses, splines and regions:
a) to a specified height and taper angle
b) along a path.

Extruded example 1: letters

1 Open your A3SOL template file, layer MODEL, Solids toolbar.

2 *a*) change viewpoint in lower left viewport to Rotate 30 and 30
   b) upper left viewport active and restore UCS RIGHT.

3 Using the reference sizes given in Fig. 29.1:
   a) draw the three letters M, T and C as polyline shapes
   b) use the polyline edit command to 'convert' the letter C into a single polyline
   c) use your discretion for sizes not given (a snap of 5 helps)
   d) use the start points A, B and C given.

4 Still with UCS RIGHT, zoom centre about 0,50,0 at 275 in all viewports.

5 Menu bar with **Draw–Solids–Extrude** and:
   prompt   Select objects
   respond **pick the letter M then right-click**
   prompt   Path/<Height of Extrusion>
   enter   **80 <R>**
   prompt   Extrusion taper angle<0> and enter: **0 <R>**.

6 The letter M will be extruded for a height of 80 in the positive *Z*-direction.

7 Select the EXTRUDE icon from the Solids toolbar and:
   prompt   Select objects
   respond **pick the letter T then right-click**
   prompt   Path/<Height of Extrusion>
   enter   **50 <R>**
   prompt   Extrusion taper angle<0> and enter: **5 <R>**.

8 At the command line enter **EXTRUDE <R>** and:
   a) objects: pick the letter C then right-click
   b) height: enter **–50**
   c) taper angle: enter **–5**.

9 Hide and shade each viewport, then save if required.

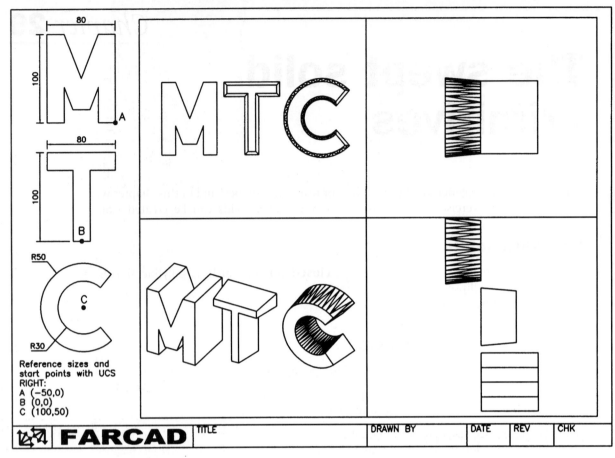

Figure 29.1 Extruded model 1 – letters.

Extruded example 2: keyed splined shaft

1 Open the A3SOL template file, layer MODEL, UCS BASE and the lower right viewport active.

2 Zoom centre about 0,0,–20 at 150 magnification in all viewports.

3 Refer to Fig. 29.2 and create two profiles:
   a) an outer tooth profile from two circles and an arrayed line, then trim as required. The circle centres should be at 0,0
   b) an inner shaft profile to your own specification.

4 Use the menu bar sequence Modify–Objects–Polyline (or PEDIT at the command line) to convert each profile into a single polyline using the Join option.

5 Select the EXTRUDE icon from the Solids toolbar and:

prompt	Select objects
respond	**pick the outer tooth profile then right-click**
prompt	Path/<Height of Extrusion>
enter	**–70 <R>**
prompt	Extrusion taper angle<0>
enter	**–3 <R>**.

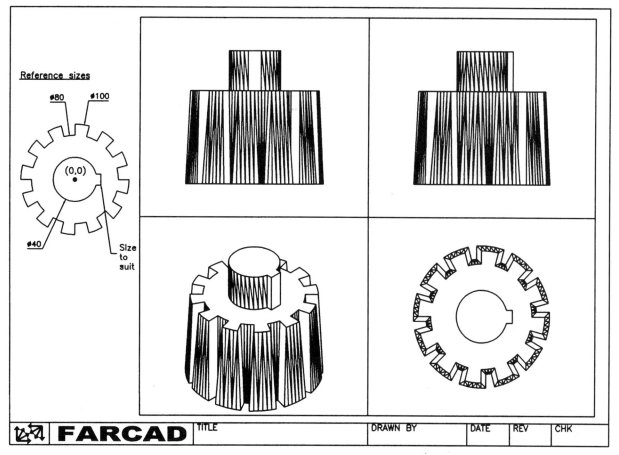

Reference sizes

Figure 29.2 Extruded model 2 – splined shaft.

6 Repeat the Extrude icon selection and:
   a) objects: pick the inner shaft profile then right-click
   b) height: enter 30
   c) taper: enter 0.

7 Hide the viewports, but do not shade – why?

8 Save the model if required.

Extruded example 3: moulding

1 Open the A3SOL template file with UCS BASE, layer MODEL and with the lower right viewport active.

2 Draw a polyline:
   from: 0,0
   to: @0,100
   arc endpoint: @100,0
   arc endpoint: @100,–100
   line endpoint: @100,0 then right-click.

3 Change the colour of the polyline to blue.

4 Restore UCS FRONT and make the top right viewport active.

5   Use the reference sizes in Fig. 29.3 to create the moulding as a single polyline – you can
    create your own design if required.

6   Make the lower left viewport active.

7   Select the EXTRUDE icon from the Solids toolbar and:

prompt	Select objects
respond	**pick the red polyline then right-click**
prompt	Path/<Height of Extrusion>
enter	**P <R>** – the path option
prompt	Select path
respond	**pick the blue polyline**
and	the red moulding is extruded along the blue path

8   Hide the viewports – shade?

9   Save if required.

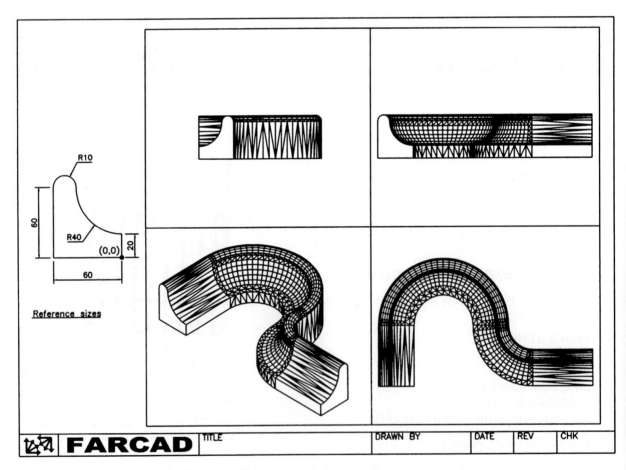

Figure 29.3   Extruded model 3 – the moulding.

Extruded example 4: a polygon duct arrangement

1 Open the A3SOL template file with UCS BASE, layer MODEL and the lower left viewport active.

2 Create a five-segment 3D polyline with the menu bar sequence **Draw–3D Polyline** and:
   prompt   From point and enter: **0,0 <R>**
   prompt   To point and enter: **@0,0,100 <R>**
   prompt   To point and enter: **@100,0,100 <R>**
   prompt   To point and enter: **@100,100,0 <R>**
   prompt   To point and enter: **@0,100,100 <R>**
   prompt   To point and enter: **@100,100,–100 <R>**
   prompt   To point and: **right-click**.

3 Change the colour of the 3D polyline to green.

4 In each viewport, zoom centre about 150,150,175 at 400 mag.

5 Draw a polygon with:
   a) sides: 6
   b) centre: 0,0
   c) inscribed in a circle of radius 40.

6 Select the EXTRUDE icon and:
   a) objects: pick the red polygon then right-click
   b) enter P <R> for the path option
   c) path: pick the green 3D polyline.

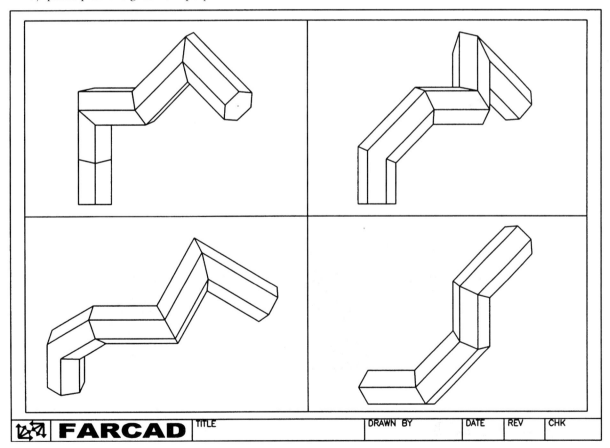

FARCAD

| | TITLE | | DRAWN BY | | DATE | REV | CHK |

Figure 29.4 Extruded model 4 – a polygon duct.

7 The polygon will be extruded along the green path to give a ducting effect.

8 Hide then save the model if required to complete the extrusion exercises.

Revolved solids

Solid models can be created by revolving objects (closed polylines, polygons, circles, ellipses, closed splines, regions) about the *X*- and *Y*-axes by a specified angle. As with extrusions, very complex models can be obtained from relatively simple shapes.

Revolved example 1: a shaft

1 Open the A3SOL template file, UCS BASE, Layer MODEL and the lower right viewport active.

2 Refer to Fig. 29.5 and draw a polyline outline using the sizes given. Use the (0,30) start point.

3 Menu bar with **Draw–Solids–Revolve** and:

prompt	`Select objects`
respond	**pick the polyline then right-click**
prompt	`Axis of revolution - Object/X/Y/<Start point of axis>`
enter	**X <R>** – the *X*-axis option
prompt	`Angle of revolution<full circle>`
respond	**right-click** to accept the full circle default.

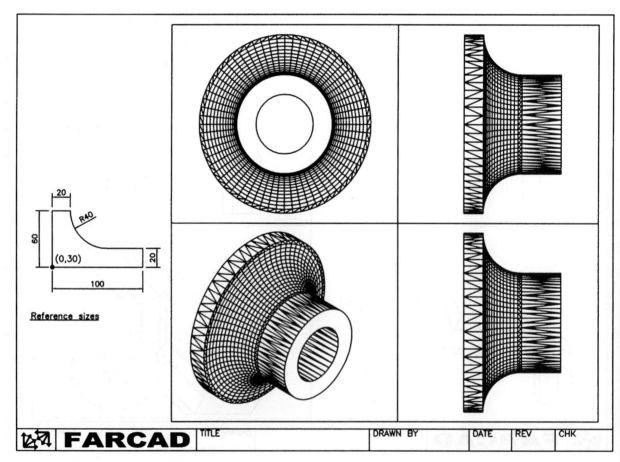

Figure 29.5 Resolved model 1 – shaft.

4   The polyline outline will be revolved about the *X*-axis.

5   Zoom centre about 50,0,0 at 200 magnification in all viewports.

6   Hide and save.

Revolved example 2: a bearing 'of sorts'

1   Open the A3SOL template file with UCS BASE, layer MODEL and the lower right viewport active.

2   Zoom centre about 0,0,0 at 300 magnification in each viewport.

3   Refer to Fig. 29.6 and draw a polyline outline using the overall sizes given. Use the (20,–20) start point and design the outline to your own specification.

4   Select the REVOLVE icon from the Solids toolbar and:
    prompt   Select objects
    respond   **pick the red outline then right-click**
    prompt   Axis of revolution – Object/X/Y...
    enter   **X <R>** – the X-axis
    prompt   Angle of revolution<full circle>
    enter   **250 <R>**.

5   Hide the viewports – Fig. 29.6(A).

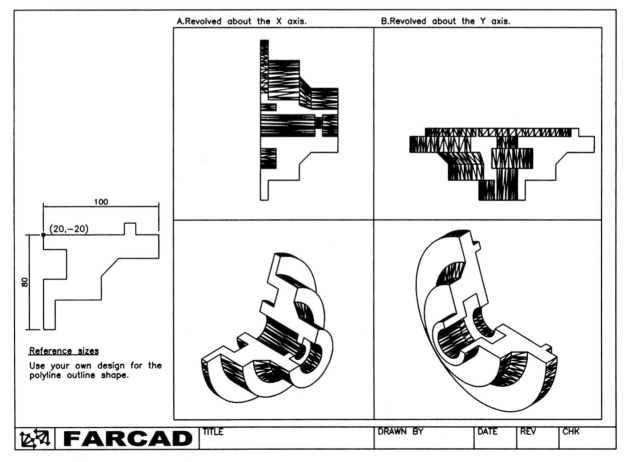

Figure 29.6   Revolved model 2 – a bearing housing.

6 *Note*: only the 3D and top viewports have been displayed.

7 Undo the hide and revolved effect (**U <R>**).

8 At the command line enter **REVOLVE <R>** and create another revolved solid with:
  a) objects: pick the red outline and right-click
  b) axis of revolution: enter **Y <R>** – the Y axis option
  c) angle of revolution: enter **250 <R>**.

9 Result is Fig. 29.6(B) – two viewports only displayed.

10 Save the model if required, we will not refer to it again.

11 This completes the swept primitive exercises.

Summary

1 Swept solids are obtained with the extrude and revolve commands.

2 The two commands can be activated:
  a) by icon selection from the Solids toolbar
  b) from the command line with Draw–Solids
  c) by entering EXTRUDE and REVOLVE at the command line.

3 Very complex models can be obtained from simple shapes.

4 Only certain 'shapes' can be extruded/revolved. These are closed polylines, circles, ellipses, polygons, closed splines and regions (more on this in a later chapter).

5 Objects can be extruded:
  a) to a specified height
  b) with/without a taper angle
  c) along a path curve.

6 The extruded height is in the Z-direction and can be positive or negative.

7 The taper angle can be positive or negative.

8 Objects can be revolved:
  a) about the X- and Y-axes
  b) about an object
  c) by specifying two points on the axis of revolution.

9 The angle of revolution can be full (360°) or partial.

Chapter 30

Boolean operations and composite solids

The basic and swept solids which have been created are called **primitives** and are the 'basic tools' for solid modelling. With these primitives the user can create **composite solids**, so called because they are 'composed' of two or more solid primitives, i.e.

a) primitive: a box, wedge, cylinder, extrusion, etc.

b) composite: a solid made from two or more primitives.

Composite solids are created from primitives using the three **Boolean** operations of union, subtraction and intersection. Figure 30.1 demonstrates these operations with two primitives:

a) a box

b) a cylinder 'penetrating' the box.

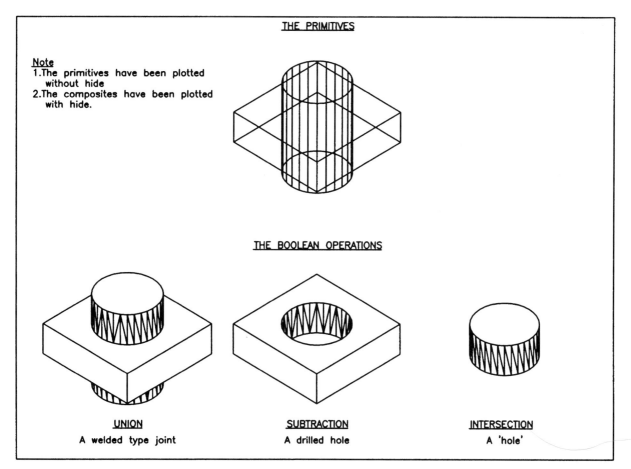

THE PRIMITIVES

Note
1. The primitives have been plotted without hide
2. The composites have been plotted with hide.

THE BOOLEAN OPERATIONS

UNION
A welded type joint

SUBTRACTION
A drilled hole

INTERSECTION
A 'hole'

Figure 30.1 The three Boolean operations.

Union

1 This operation involves 'joining' two or more primitives to form a single composite, the user selecting all objects to be unioned.

2 The operation can be considered similar to welding two or more components together.

Subtraction

1 This involves removing one or more solids from another solid thereby creating the composite. The user selects:
   a) the source solid
   b) the solids to be subtracted from the source solid.

2 The result of a subtraction operation can be likened to a drilled hole, i.e. if the cylinder is subtracted from the box, a hole will obtained in the box.

Intersection

1 This operation gives a composite solid from other solids which have a common volume, the user selecting all objects which have to intersect.

2 The box/cylinder illustration of the intersection operation gives a 'disk shape' or 'hole', i.e. if the box and cylinder are intersected, the common volume is the disk shape.

Creating a composite solid from primitives

There is no 'correct or ideal' method of creating a composite, i.e. the Boolean operations selected by one user may be different from those selected by another user, but the final composite may be the same. To demonstrate this, we will create an L-shaped component by three different methods, so:

1 Open your A3SOL template file and:
   a) enter paper space
   b) erase any text and the four viewports
   c) with layer VP current create a single viewport with:
       i) first point: pick to suit in lower left corner area
       ii) other corner: enter @360,250.

2 Return to model space, UCS BASE, layer MODEL.

3 Create two box primitives:
   a) corner: 0,0,0
       cube option
       length: 100
       colour: red

   b) corner: 100,100,100
       length: −60
       width: −100
       height: −70
       colour: blue

4 Create another two box primitives:
   a) corner: 125,125,0
       length: 100
       width: 100
       height: 30
       colour: red

   b) corner: 125,125,30
       length: 40
       width: 100
       height: 70
       colour: blue

5 Restore UCS FRONT and draw a 2D polyline shape:
   from   500,250   *to*   @100,0   *to*   @0,30   *to*   @−60,0
   to   @0,70   *to*   @−40,0   *to*   close

6 Restore UCS BASE and zoom centre about 100,250,0 at 400 mag.

7 From the menu bar select **Modify–Boolean–Subtract** and:
   prompt   Select solids and regions to subtract from...
              Select objects
   respond **pick large left red box then right-click**
   prompt   Select solids and regions to subtract...
              Select objects
   respond **pick left blue box then right-click**
   and   the blue box is subtracted from the red box.

8 Menu bar with **Modify–Boolean–Union** and:
   prompt   Select objects
   respond **pick the middle red and blue boxes then right-click**
   and   the two boxes will be unioned.

9 Restore UCS FRONT and:
   a) select the EXTRUDE icon from the Solids toolbar
   b) pick the L shaped polyline then right-click
   c) enter an extruded height of −100 with 0 taper
   d) the L-shape polyline is extruded into a composite.

10 *Task.*
   a) Hide the models – all the same
   b) Shade the models – note colour effect between the union and subtraction composites
       – any comment?
   c) At the command line enter **MASSPROP <R>** and:
       prompt   Select objects
       respond **pick left composite then right-click**
       and   AutoCAD Text Window with:
                  Mass: 580000.00
                  Volume: 580000.00
                  Bounding box, Centroid, etc.
       enter   **N <R>** in response to 'write to file' prompt
   d) Repeat the MASSPROP command and select the middle and right composites – same
       mass and volume?

11 *Question.*
   a) Why are the mass and volume the same?
       Answer: Release 14 assumes a density value of 1 and does not support a materials
       library as Release 12.
   b) Is the volume of 580000 correct for the L shape?
   c) What are the volume units?

12 Now that we have investigated the Boolean operations, we will create some composite
   solid models which (I hope) will be interesting.

13 The three methods are displayed in Fig. 30.2.

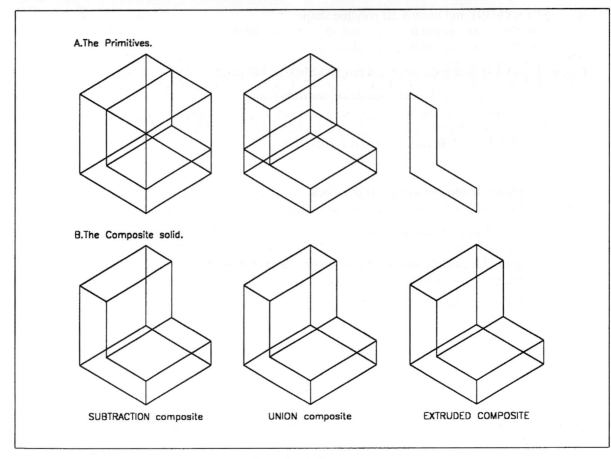

Figure 30.2 Creating a composite by three different methods.

Summary

1 There are three Boolean operations – union, subtraction and intersection.

2 The three operations be activated:
   a) from the menu bar with Modify–Boolean
   b) in icon form from the Modify II toolbar
   c) by entering the command from the keyboard.

Chapter 31

Composite model 1 – a machine support

In this exercise we will create a composite solid from the box, wedge, cylinder and cone primitives using the three Boolean operations. Once created, we will dimension the model using viewport-specific layers.

The exercise is quite simple and you should have no difficulty in following the various steps in the model construction. Try and work out why the various entries are given – do not just accept them.

1 Open your A3SOL template file with UCS BASE, layer MODEL and with the lower left viewport active. Display the Solids, Modify II and other toolbars to suit.

2 Refer to Fig. 31.1 which displays the 3D viewport of the model at various stages of its construction.

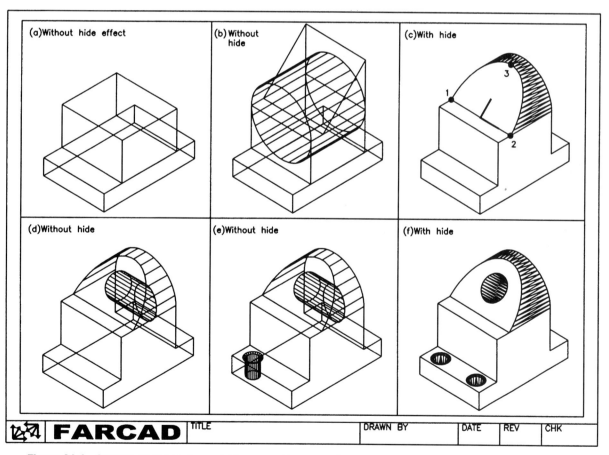

Figure 31.1 Creation of composite model 1 – a machine support.

3 Zoom centre about 50,75,75 at 225 magnification in all viewports.

4 Using the BOX icon to create two primitives:

   a) corner: 0,0,0   *b*) corner: 100,120,25
      length: 100   length: −100
      width: 150   width: −90
      height: 25   height: 60
      colour: red   colour: blue.

5 This gives fig. (a).

6 Create a cylinder on top of the blue box with:

   a) centre: 50,30,85
   b) radius: 50
   c) height/other end: enter **C <R>** then **@0,90,0 <R>**
   d) colour: green.

7 Create a wedge with:

   a) corner: 0,120,85
   b) length: 70; width: 100; height: 70
   c) colour: magenta.

8 Rotate the magenta wedge:

   a) about the point: 0,120,85
   b) by: −90° − fig. (b).

9 Select the INTERSECTION icon from the Modify II toolbar and:

   prompt   Select objects
   respond   **pick the green cylinder and magenta wedge then right-click**.

10 Select the UNION icon from the Modify II toolbar and:

   prompt   Select objects
   respond   **pick the red and blue boxes and the intersected wedge/cylinder then right-click**.

11 The model now appears as fig. (c).

12 Refer to fig. (c) and menu bar with **Tools–UCS–3 Point** and set as follows:

   a) origin: MIDpoint icon and pick line 12
   b) *X* axis: ENDpoint icon and pick pt2
   c) *Y* axis: QUADrant icon and pick pt3 on curve.

13 The UCS icon will move and align itself on the sloped surface.

   Note: if icon does not move, **View–Display–UCS–On/Origin**.

14 Save this UCS position as SLOPE.

15 Create a cylinder with:

   a) centre: 0,35,0
   b) radius: 18
   c) height: −100
   d) colour: number 54.

16 Select the SUBTRACT icon from the Modify II toolbar and:

   prompt   Select solids and regions to subtract from...
                  Select objects
   respond   **pick the composite model then right-click**
   prompt   Select solids or regions to subtract
                  Select objects
   respond   **pick the cylinder then right-click** − fig. (d).

17 Restore UCS BASE.

18 Create the following two primitives:

    a) Cylinder   *b*) Cone

       centre: 20,15,0   centre: 20,15,25

       radius: 9   radius: 12

       height: 25   height: −5

       colour: number 174   colour: number 174.

19 Union the cylinder and cone – fig. (e).

20 Multiple copy the unioned cylinder/cone:

    a) from: 20,15

    b) by: @60,0 and by: @30,120.

21 Using the SUBTRACT icon:

    a) select the original composite then right-click

    b) pick the three cylinder/cones then right-click.

22 The model is now complete and HIDE and SHADE can be used – fig. (e).

23 Remember to REGENALL.

24 At this stage save the composite as R14MOD\MACHSUPP.

Making the viewport-specific layers

The screen displays the model in a four viewport configuration and we now want to add some dimensions. These dimensions must be added on viewport-specific layers and these layers must now be created.

1 At the command line enter **VPLAYER <R>** and:

    prompt   ?/Freeze/Thaw/reset/Newfrz/Vpvisdflt

    enter   **N <R>** – the Newfrz (new viewport freeze) option

    prompt   New viewport frozen layer name(s)

    enter   **DIMTL,DIMTR,DIMBR <R>**

    prompt   ?/Freeze/Thaw/...

    respond   **right-click** – as finished with command.

2 The command line entry VPLAYER is for viewport layer.

3 Make the top left viewport active and menu bar with **Format–Layer** and:

    prompt   Layer and Linetype Properties dialogue box

    note   three new layers – Dimbr, Dimtl, Dimtr with:

                  i) Frozen in current viewport – blue icon

                  ii) Frozen in new viewport – blue icon

    respond   1. **pick Dimtl** (turns blue) and note the details – Freeze in current viewport

                   is on (tick)

           2. **pick blue icon Freeze in current viewport** to Thaw the layer and tick

                 removed from Freeze in current viewport details list

           3. pick OK.

4 With the top right viewport active, Format–Layer and:

    a) pick layer **Dimtr**

    b) toggle the blue Freeze in current viewport icon to yellow to Thaw layer Dimtr in the

        top right viewport

    c) pick OK.

5 With lower right viewport active, Format–Layer and:
   a) pick layer **Dimbr**
   b) toggle Freeze icon in current viewport from blue to yellow, i.e. from Frozen to Thaw to Thaw layer Dimbr in the lower right viewport
   c) pick OK.

6 *What has been achieved in this section?*
   a) Three new viewport specific layers have been made.
   b) These layers have been named DIMTL for the top left viewport, DIMTR for the top right viewport and DIMBR for the bottom left viewport.
   c) The three layers were originally created:
       i) frozen in new viewports
       ii) currently frozen in all viewports.
   d) Each layer was currently thawed in a specific viewport, i.e. layer Dimtl is currently thawed in the top left viewport but is currently frozen in the other three viewports. Layers Dimtr and Dimbr are currently frozen in the top left viewport.

7 Before adding the dimensions, change the colour of the three new layers (Dimtl, Dimtr, Dimbr) to magenta with Format–Layer.

Adding the dimensions

1 Before the dimensions are added to the model, menu bar with **Dimension–Style** and using Dimension Styles dialogue box:
   a) select Geometry
   b) alter Overall Scale to 2 then pick OK
   c) pick Save then OK
   d) this will scale all dimensions by 2.

2 Make the lower right viewport active and:
   a) restore UCS BASE
   b) make layer Dimbr current.

3 Display the Dimension toolbar and position to suit and refer to Fig. 31.2.

4 Select the LINEAR dimension icon and add:
   a) the horizontal dimension
   b) the four vertical dimensions using the baseline option.

5 With the top right viewport active:
   a) restore UCS FRONT
   b) make layer Dimtr current
   c) add the three linear and the radius dimensions.

6 Make the top left viewport active and:
   a) restore UCS RIGHT
   b) make layer Dimtl current
   c) add the six dimensions.

7 The composite model is now complete with dimensions added and can be plotted with the VP layer frozen for effect – Fig. 31.2.

8 These dimension additions do not need to be saved.

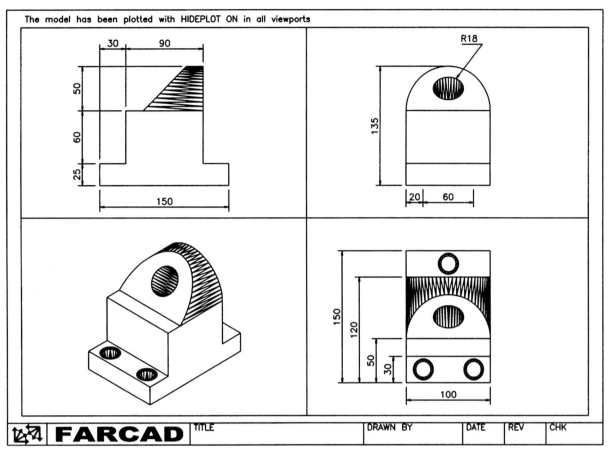

Figure 31.2 Completed solid model MACHINE SUPPORT with dimensions added.

Summary

1 Primitives and the Boolean operations can be used to create composite solid models.

2 Viewport specific layers are essential when adding dimensions to a multi-view model.

Chapter 32

Composite model 2 – a backing plate

In this exercise we will create a solid from an extruded swept primitive and then subtract various 'holes' to complete the composite. The exercise will also involve altering the viewport layout of the A3SOL template file which is an interesting exercise in itself. As with all the exercises, do not just accept the entries – work out why the various values are being used.

The model

Refer to Fig. 32.1 which details the model to be created and gives the relevant sizes. As an aside draw the three orthogonal views as given and then add the isometric (the arc 'hole' is interesting to complete in an isometric view). Time how long it takes to complete this 2D drawing. I took about an hour to complete the four views with dimensions.

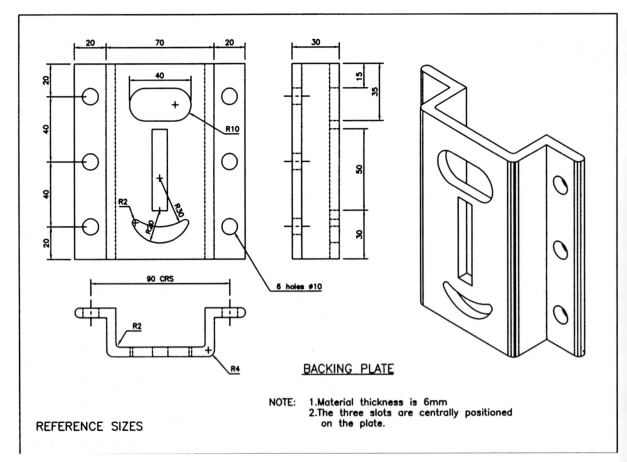

REFERENCE SIZES

BACKING PLATE

NOTE: 1. Material thickness is 6mm
2. The three slots are centrally positioned on the plate.

Figure 32.1 Backing plate drawn as orthogonal views and an isometric.

Setting the viewports

1 Open your A3SOL template file and:
   a) enter paper space
   b) erase the four viewports
   c) make layer VP current
   d) refer to Fig. 32.2.

2 Menu bar with **View–Floating Viewports–1 Viewport** and:
   prompt   First point and enter: **10,25 <R>**
   prompt   Other corner and enter: **@165,200 <R>**.

3 Create another three single viewports using the following coordinate entries:
   a) first point: 175,25
      other corner: @165,70
   b) first point: 175,95
      other corner: @165,195
   c) first point: 340,95
      other corner: @70,195.

4 In model space:
   a) make layer MODEL current
   b) restore UCS BASE
   c) set the 3D viewpoints in the viewports as fig. (a)
   d) zoom centre about 0,10,60 at **1XP** – yes enter 1XP
   e) make viewport B active.

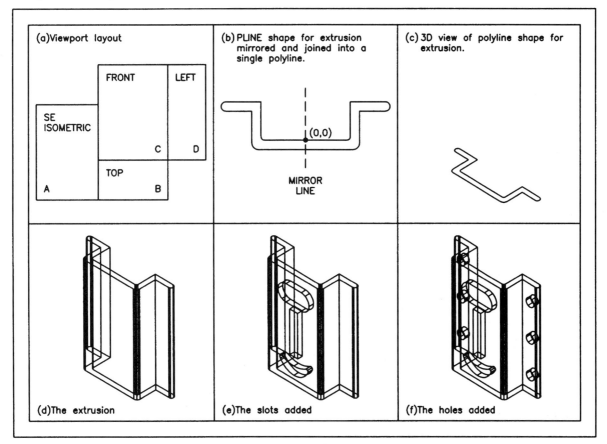

Figure 32.2 Steps in the creation of the backing plate composite.

Creating the extrusion

1 Using the polyline icon from the Draw toolbar, create a single polyline from line and arc segments with the following entries:

From point	0,0
To point	@27,0
To point	Arc option to @2,2
To point	Line option to @0,22
To point	@23,0
To point	Arc option to @0,–6
To point	Line option to @–17,0
To point	@0,–20
To point	Arc option to @–4,–4
To point	Line option to @–31,0
To point	right-click.

2 Mirror the polyline shape about a vertical line through 0,0.

3 Use the menu bar sequence **Modify–Object–Polyline** and:
   a) pick any point on the right-hand polyline
   b) enter **J <R>** – the join option
   c) pick the left-hand polyline then right-click
   d) enter **X <R>** to end the sequence.

4 The two 'halves' of the polyline have been joined into a single polyline as fig(b) in plan view and fig. (c) in 3D.

5 At the command line enter **ISOLINES <R>** and:

prompt	New value for ISOLINES<24>
enter	**6 <R>**
Note:	the ISOLINES system variable has been reduced from 24 to 6 due to the 'corners' of the model. With a value of 24, the extrusion would result in these corners being 'very dense'.

6 Select the EXTRUDE icon from the Solids toolbar and:

prompt	Select objects
respond	**pick any point on the polyline then right-click**
prompt	Path/<Height of extrusion> and enter: **120 <R>**
prompt	Extrusion taper angle and enter: **0 <R>**.

7 The polyline shape will be extruded as fig. (d).

Creating the 'holes'

1 Restore UCS FRONT with viewport C active.

2 Create a box primitive with:
   a) corner: –5,30,0
   b) length: 10; width: 50; height: 6
   c) colour: blue.

3 Create the following three primitives:

box	*cylinder*	*cylinder*
corner: –10,85,0	centre: –10,95,0	centre: 10,95,0
length: 20	radius: 10	radius: 10
width: 20	height: 6	height: 6
height: 6	colour: green	colour: green
colour: green.		

4 Union the three green primitives.

5 Draw two circles:
   a) centre: 0,30 with radius: 20
   b) centre: 0,50 with radius: 50
   c) both circles colour green.

6 Trim the circles 'to each other' and fillet the 'corners' with a radius of 2.

7 Convert the four arcs into a single polyline with the menu bar sequence **Modify–Object–Polyline** using the **J**oin option.

8 Extrude the green polyline for a height of 6 with 0 taper.

9 Subtract the green and blue 'holes' from the red extrusion to display the model as fig. (e).

10 Create a cylinder with:
   a) centre: 45,20,–24
   b) radius: 5
   c) height: 6
   d) colour: magenta.

11 Rectangular array the magenta cylinder:
   a) for 3 rows and 2 columns
   b) row distance: 40
   c) column distance: –90.

12 Subtract the six magenta cylinders from the composite.

13 The model is now complete – fig. (f) and the four viewport layout should be displayed as Fig. 32.3.

14 Hide, shade, regenall then save the model as R14MOD\BACKPLT.

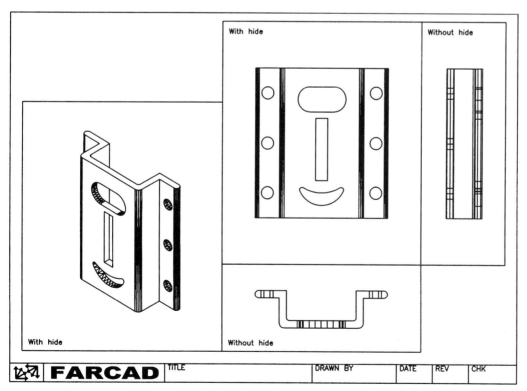

Figure 32.3 Complete solid model composite of backing plate.

Investigating the model

Display the Inquiry toolbar and make the 3D viewport active.

1 Select the LIST icon from the Inquiry toolbar and:
   prompt   Select objects
   respond   **pick the composite then right-click**
   and   AutoCAD Text Window
   with   details of the model, e.g. layer on which it was created, the space and the
              bounding box – note the coordinates of the bounding box
   then   cancel the text window.

2 Select the AREA icon from the Inquiry toolbar and:
   prompt   <First point>/Object/Add/Subtract
   enter   **0 <R>** – the object option
   prompt   Select objects
   respond   **pick the composite**
   and   Area = 38997.79, Perimeter = 0.00.
   This is the surface area of the model.

3 Select the MASS PROPERTIES icon and:
   prompt   Select objects
   respond   **pick the composite then right-click**
   prompt   AutoCAD Text Window
   with   Mass = 1002377.77 and other 'technical' information about the model
   enter   **<R>** for more information
   then   **N <R>**, i.e. do not write to file. More on this in a later chapter.

4 This completes the exercise on the extruded model.

Chapter 33

Composite model 3
– a flange and pipe

This exercise will involve creating a composite mainly as a revolved swept primitive. Cylinder primitives will be subtracted from the revolved primitive to complete the composite.

1 Open A3SOL template file with layer MODEL and UCS BASE. Display the Solids toolbar.

2 Zoom centre about −100,−30,0 at 400 magnification in all viewports.

3 With the lower left viewport active, restore UCS RIGHT and draw:
   a) circle, centre: 0,0 with radius: 30
   b) circle, centre: 0,0 with radius: 40
   c) line, *from* 0,0 *to* @−200,0 *to* @0,100.

4 Select the REVOLVE icon from the Solids toolbar and:
   prompt   Select objects
   respond   **pick the smaller circle then right-click**
   prompt   Axis of revolution - Object/X/Y...
   enter   **O <R>** – the object option
   prompt   Select an object
   respond   **pick the lower end of the vertical line**
   prompt   Angle of revolution
   enter   **70 <R>**.

5 Change the colour of the revolved pipe to green.

6 Revolve the larger circle using the same entries as step 4 and change the colour of the pipe to blue.

7 Erase the two lines, then subtract the green pipe from the blue pipe – may need a paper space zoom for this operation?

8 Make the lower right viewport active and restore UCS BASE.

9 With the POLYLINE icon from the Draw toolbar, draw a continuous polyline with the following entries:
   From point   0,−30
   To point   @0,−10
   To point   @10,0
   To point   Arc option with endpoint: @10,−10
   To point   Line option with endpoint: @0,−50
   To point   @30,0
   To point   @0,70
   To point   close option.

10 Using the REVOLVE icon from the Solids toolbar:
    a) pick a point on the polyline then right-click
    b) enter **X <R>** as the axis of revolution
    c) enter **360 <R>** as the angle.

11 Change the colour of the revolved solid to magenta.

12 With the top left viewport active, restore USC RIGHT.

13 Create a cylinder with:
    a) centre: 75,0,50
    b) radius: 10
    c) height: −30
    d) colour: number 10.

14 Polar array this cylinder about the point 0,0 for six items.

15 *a*) subtract the six cylinders from the magenta flange
    b) union the flange and pipe.

16 The model is complete and can be saved as R14MOD\FLPIPE.

17 Hide, shade, etc. before leaving the exercise.

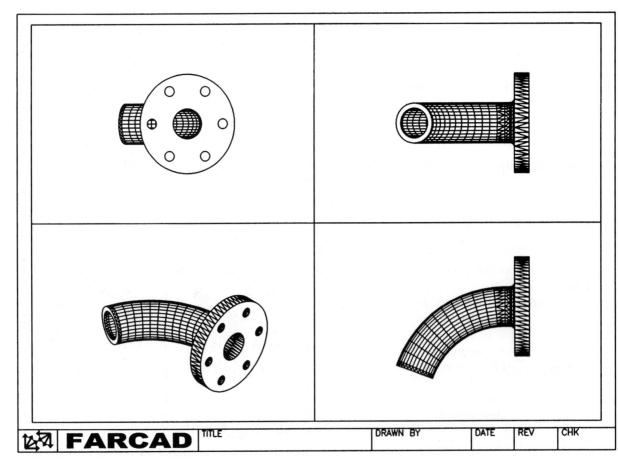

Figure 33.1 Creation of composite model 3 – pipe/flange.

The edge primitives

Models can be modified to include a chamfer and fillet effect. These are the edge primitives – the third type of primitive which can be created. In this chapter we will investigate how solids can be constructed with these edge primitives.

Example 1 – a box solid

1 Open the A3SOL template file with the lower left viewport active, UCS BASE and layer MODEL. Display the Solids toolbar.

2 Refer to Fig. 34.1(a).

3 Use the BOX icon to create a cube with:
   a) corner: 0,0,0
   b) length: 100.

4 Zoom centre about 50,50,50 at 225 magnification.

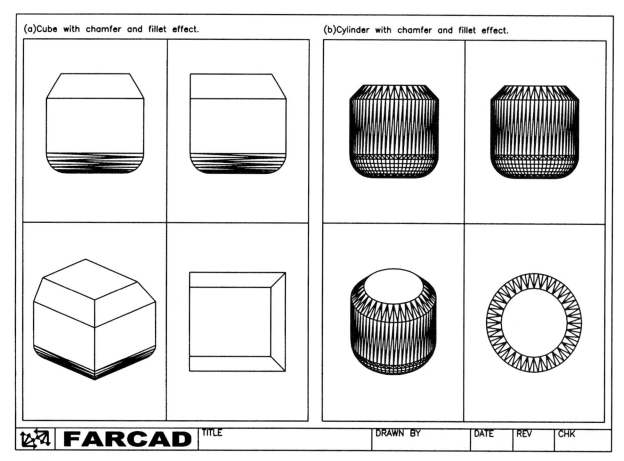

Figure 34.1 The EDGE primitives – modifying solids.

5 Select the CHAMFER icon from the Modify toolbar and:

prompt	`Polyline/Distance...`
respond	**pick any line on top surface**
prompt	`Select base surface`
and	*a*) one face of the cube will be highlighted
	b) it will be a 'side' or the 'top'
	c) prompt is Next/<OK>
respond	*a*) right-click if top face is highlighted
	b) enter N<R> and top face highlighted then right-click
prompt	`Enter base surface distance<?>`
enter	**15 <R>**
prompt	`Enter other surface distance<?>`
enter	**25 <R>**
prompt	`Loop/<Select edge>`
respond	*a*) pick any three sides on top surface
	b) right-click
and	the top surface will be chamfered at the selected three edges

Note: entering **L** for loop will allow all edges to be chamfered with a single pick.

6 Select the FILLET icon from the Modify toolbar and:

prompt	`Polyline/Radius/...`
respond	**pick any line on the base surface of the cube**
prompt	`Enter radius<?>`
enter	**20 <R>**
prompt	`Chain/Radius/<Select edge>`
enter	**C <R>** – the chain option
prompt	`Chain/Radius/<Select edge chain>`
respond	**pick the four edges of the base then right-click**
and	the cube base is filleted.

7 The cube is now displayed as fig. (a).

Example 2 – a cylinder solid

1 Erase the cube and at the command line enter **ISOLINES <R>** and check the value is 24.

2 Use the CYLINDER icon to create a cylinder with:
   a) centre: 0,0,0
   b) radius: 50
   c) height: 100.

3 Zoom centre about 0,0,50 at 200 magnification.

4 In paper space zoom-in on the 3D viewport then return to model space.

5 Select the CHAMFER icon and:

prompt	`Polyline/Distance/...`
respond	**pick the top surface circle edge**
prompt	`Select base surface`
and	`Next/<OK>`
respond	**right-click** as the required surface is highlighted
prompt	`Enter base surface distance` and enter: **15 <R>**
prompt	`Enter other surface distance` and enter: **15 <R>**
prompt	`Loop/<Select edge>`
respond	**pick top circle edge then right-click**.

6   The top of the cylinder is chamfered with the entered values.

7   Select the FILLET icon and:
     prompt   Polyline/Radius/...
     respond   **pick bottom circle of cylinder**
     prompt   Enter radius
     enter   **25 <R>**
     prompt   Chain/Radius/<Select edge>
     respond   **right-click** as bottom edge already selected.

8   The cylinder is filleted at the base.

9   In paper space zoom-previous then return to model space.

10 The chamfer/fillet effect of the cylinder is displayed as fig. (b).

A composite edge primitive solid

1   Erase the cylinder model and with UCS BASE, layer MODEL and the lower left viewport active, create three primitives with:

box	*cylinder*	*box*
corner: 0,0,0	centre: 75,60,0	corner: 50,0,40
length: 150	radius: 40	length: 60
width: 120	height: 100	width: 120
height: 100	colour: green	height: 40
colour: red		colour: blue

2   In all viewports, zoom centre about 75,60,50 at 200 mag.

3   Subtract the green and blue primitives from the red box and note the 'interpenetration' effect.

4   Select the FILLET icon and:
     a) pick the top circle of the green object
     b) enter a radius of 15
     c) right-click.

5   The top edge of the cylinder is filleted 'outwards' and is red. Why is the fillet red and not green?

6   Select the CHAMFER icon and:
     a) pick the top long front edge of the red box
     b) right-click if the front vertical face is highlighted or N <R> until front vertical face is highlighted then right-click
     c) enter base surface distance of 10
     d) enter other surface distance of 5
     e) pick the four front edges of the blue primitive and right-click.

7   The blue box primitive is chamfered in red – not blue?

8   The model is displayed as Fig. 34.2. Hide, shade, etc.

9   The model can be saved if required, but will not be used again.

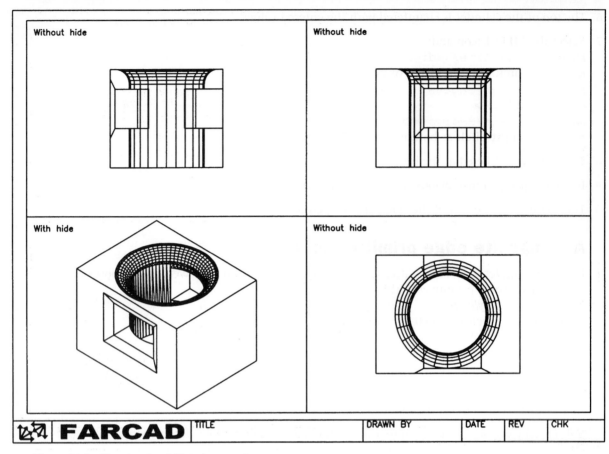

Figure 34.2 Chamfered and filleted composite.

An interesting variation?

Before leaving this chapter, we will investigate the effect of the fillet and chamfer edge primitives 'at three corners', so:

1 Erase the composite on the screen and ensure UCS BASE, layer MODEL and lower left viewport active. Refer to Fig. 34.3.

2 Draw a polygon with six sides, centre point at 50,50 and circumscribed in a circle of radius 50.

3 Extrude the hexagon for a height of 50 with 0 taper.

4 Select the FILLET icon and:
   prompt   Select first object
   respond **pick edge 23**
   prompt   Enter radius and enter: **25 <R>**
   prompt   Chain/Radius/<Select edges>
   respond **pick edges 12,23,24 then right-click**.

5 The model will be filleted at the three selected edges with a 'blend' at the corner.

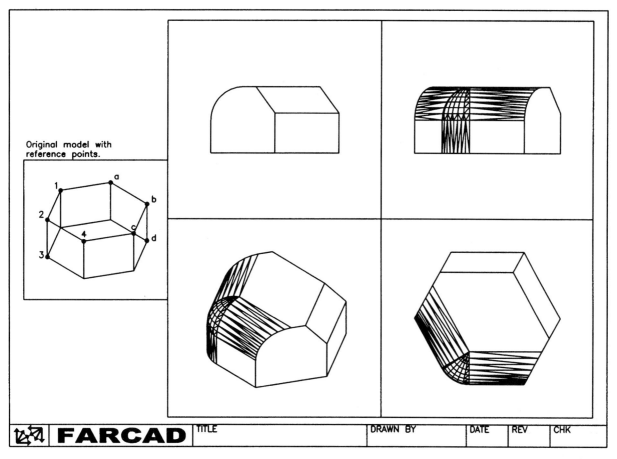

Figure 34.3 Fillet and chamfer effect at three corners.

6 Select the CHAMFER icon and:

    prompt   Select first line

    respond   **pick edge bc**

    prompt   Select base surface

                Next/<OK>

    respond   **enter N <R> until top surface is highlighted then right-click**

    prompt   Enter base surface distance and enter: **15 <R>**

    prompt   Enter other surface distance and enter: **20 <R>**

    prompt   Loop/<Select edge>

    respond   **pick edges ab and bc then right-click**.

7 The model will be chamfered at the selected corner.

Summary

1 Primitives and solids can be chamfered and filleted with the 'normal' CHAMFER and FILLET commands.

2 Solids and primitives can be chamfered/filleted:
   a) inwards if a primitive
   b) outwards if a 'hole'.

3 Individual edges can be chamfered and filleted.

4 The chamfer command has a LOOP option allowing a complete surface to be chamfered.

5 The fillet command has a CHAIN option allowing a complete surface to be filleted.

6 Error messages will be displayed if the chamfer distances or the fillet radius are too large for the model being modified, and the command line will display:
   a) failed to perform blend
   b) failure while chamfering/Filleting.

Assignment

This activity requires a cube to be chamfered to give a 'truncated pyramid'. The model will be used in a later activity.

Activity 19: Penetrated pyramid

The composite has to be created from three primitives:
a) red cube of side 200
b) green cylinder with radius 25 with the centre 100 from the base
c) blue square box of side 80 with the lower edge 30 from the base.

Suggested approach:

1 Position the cube with corner at 0,0,0.

2 Chamfer the cube to give a square topped pyramid, the top having a side length of 100.

3 Position the square sided box.

4 Position the cylinder – use other end option.

5 Subtract the box and cylinder from the pyramid.

6 Chamfer the blue square box 'front', distances: 10.

7 Fillet the green cylinder 'right side' with radius 10.

8 Save the composite as R14MOD\MODCOMP.

Chapter 35

Four solid model composite exercises

In this chapter four solid model composites will be created to 'reinforce' the process of how composites are created from primitives. Hopefully you will enjoy creating the models. Each model will use the A3SOL template file with layer MODEL. The models can be saved if required.

Interpenetration model – Fig. 35.1

1   With UCS BASE, draw a polygon with:
   a) sides: 6
   b) centre: 0, 0
   c) circumscribed in circle of radius: 50.

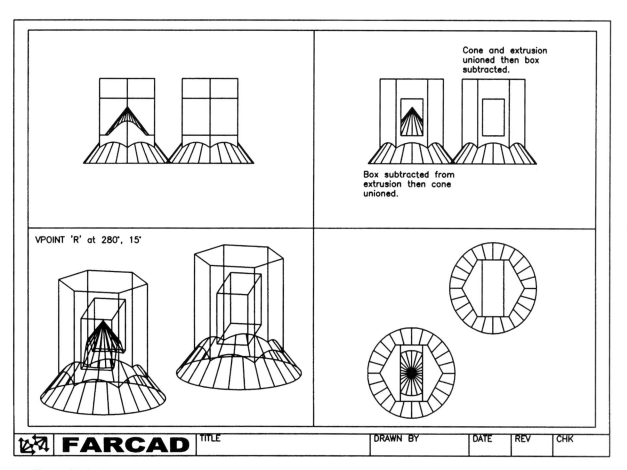

Figure 35.1   Interpenetration exercise – model plotted without hide effect.

2 Extrude the hexagon to a height of 150 with 0 taper.

3 Create two primitives:

cone	box
centre: 0,0,0	corner: −20, −50, 50
radius: 80	length: 40
height: 100	width: 100
colour: green	height: 65
	colour: blue.

4 Copy the three primitives from: 0, 0 by: @150, 150, 0.

5 Zoom centre about 80, 80, 60 at 350 magnification.

6 With the left hand model:
   a) subtract the blue box from the red extrusion
   b) union the green cone and the red extrusion.

7 With the right hand model:
   a) union the green cone and red extrusion
   b) subtract the blue box from the composite.

8 Note the cone effect 'in the box' – the order of the Boolean operations is important.

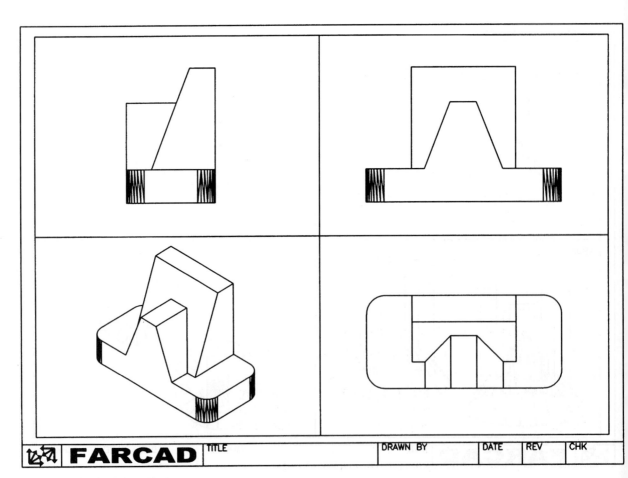

Figure 35.2 Bolster block.

Bolster block – Fig. 35.2

1 With UCS BASE create a box with:
   a) corner: 0, 0, 0
   b) length: 150; width: 70; height: 25.

2 Zoom centre about 75, 40, 50 at 150 magnification.

3 Fillet the four 'edges' of the box with radius 14.

4 Create two primitives:
box	*wedge*
corner: 35, 70, 25	corner: 115, 50, 25
length: 80	length: 30
width: –20	width: –80
height: 76	height: 76

5 Rotate the wedge about 115,50 by –90 degrees.

6 Create another two primitives:
box	*wedge*
corner: 65, 0, 25	corner: 85, 0, 25
length: 20	length: 20
width: 60	width: 60
height: 50	height: 50

7 Mirror the wedge about the points 75, 0 and 75, 100.

8 Union all the primitives.

Fulcrum support – Fig. 35.3

1 With top right viewport active restore UCS FRONT and draw a polyline using:
from	0, 0
to	@0, 28
to	arc option with endpoint @4, 4
to	line option with endpoint @92, 0
to	arc option with endpoint: @4, –4
to	line option with endpoint: @0, –38
to	@–16, 0
to	@0, 22
to	arc option with endpoint @–4, 4
to	line option with endpoint @–60, 0
to	arc option with endpoint @–4, –4
to	line option with endpoint @0, –12
to	close.

2 Extrude the polyline for a height of –64 with 0 taper.

3 Restore UCS BASE and zoom centre about 50, 32, 0 at 150 mag.

4 Create three cylinders with:
a) centre: 50, 32, 16	*b*) centre: 50, 32, 32	*c*) centre: 50, 32, 52
radius: 8	radius: 25	radius: 15
height: 36	height: 20	height: –5.

5 *a*) Union the largest cylinder and the extrusion
   b) subtract the two smaller cylinders from the composite.

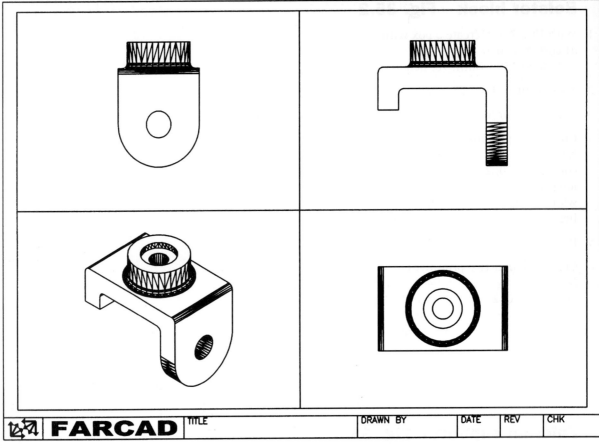

Figure 35.3 Fulcrum support.

6 Fillet the largest cylinder and extrusion 'join' with radius 4.

7 Restore UCS RIGHT and create two cylinders:
    a) centre: 32, −10, 100   *b*) centre: 32, −10, 100
       radius: 32   radius: 10
       height: −16   height: −16.

8 Union the larger cylinder and the composite, then subtract the smaller cylinder from the composite.

9 Restore UCS BASE.

Arched sculpture – Fig. 35.4

1 Restore UCS RIGHT and draw a polyline:
   From   0,0   *To*   @0,100   *To*   @50,50   *To*   @0,40
   To   arc endpoint @50,50
   To   line endpoint @50,0
   To   right click.

2 Restore UCS BASE and draw another polyline:
   From   20,−20   *To*   @0,40   *To*   @−40,0
   To   @0,−40   *To*   close.

3 Extrude the square polyline along the polyline path and erase the polyline path.

4 Zoom centre about 0, 185, 150 at 500 magnification.

5 Draw a polygon with:
  a) sides: 6
  b) edge option
  c) first point of edge: −20, 150, 230
  d) second point of edge: @40, 0.

6 Extrude the hexagon for a height of 100 with 0 taper and change its colour to blue.

7 Polar array the red extruded component:
  a) about the point: 0,185
  b) for six items
  c) full circle with rotation.

8 Union all the primitives.

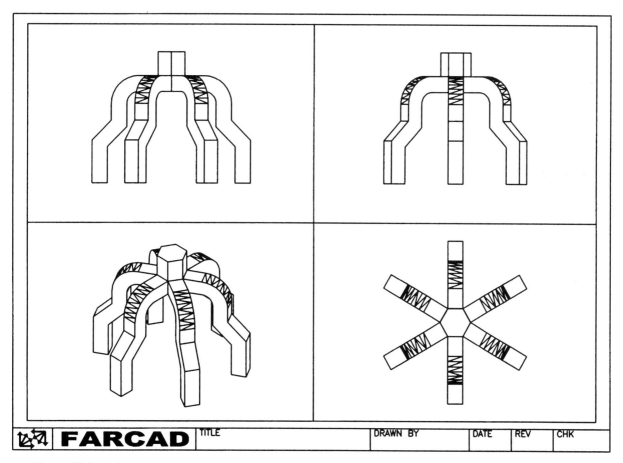

Figure 35.4 Arched sculpture.

Chapter 36

Regions

A region is a closed 2D shape created from lines, circles, arcs, polylines, splines, etc. and can be used with the extrude and revolve command to create solid composites. When created, a region has certain characteristics:

- it is a solid of zero thickness
- it is coplaner, i.e. must be created on the one plane
- it consists of loops – outer and inner
- the loops must be continuous closed shapes
- every region has one outer loop
- there may be several inner loops
- inner loops must be in the same plane as the outer loop
- regions can be created with the Boundary command
- regions can be used with EXTRUDE or REVOLVE.

With the exception of the boundary option, using regions does not allow the user any new ability to create solid models – it is another variation to be considered. We will investigate regions with several worked examples.

Example 1 – the letter C

1   Open your A3SOL template file with layer MODEL, UCS BASE and the lower right viewport active. Refer to Fig. 36.1 which only displays the 3D viewport of the exercise.

2   Create the letter C from two circles and two lines using the sizes given. Trim the lines and circles.

3   Using the Edit Polyline command with the Join option, convert the four objects into a single polyline – fig. (a).

4   Zoom centre about 0,0,40 at 200 mag in all viewports.

5   With the lower left viewport active, select the REGION icon from the Draw toolbar and:
   a) pick the letter then right-click
   b) enter an extruded height of 75
   c) enter a taper angle of 0
   d) the letter is extruded as fig. (b)
   e) hide the model.

6   Undo the hide and extrusion effect with **U <R>** and **U <R>**.

7   Select the EXTRUDE icon again and:
   a) pick the letter then right-click
   b) enter a height of 75
   c) enter a taper angle of 8
   d) result is fig. (c) with hide.

8   Undo the hide and extrusion effect.

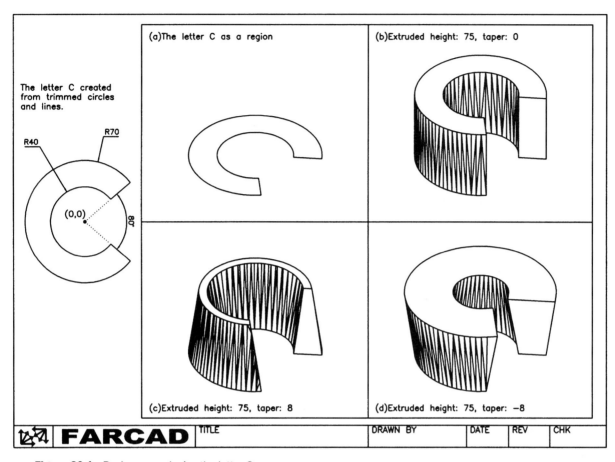

The letter C created from trimmed circles and lines.

R40 R70

(0,0)

80

(a)The letter C as a region

(b)Extruded height: 75, taper: 0

(c)Extruded height: 75, taper: 8

(d)Extruded height: 75, taper: −8

FARCAD | TITLE | | DRAWN BY | DATE | REV | CHK

Figure 36.1 Region example 1 – the letter C.

9 Extrude the letter:
   a) for a height of 75
   b) with a taper angle of −8
   c) result is fig. (d).

10 Save the exercise if required.

11 *Note*: a possible error message will occur if the taper angle is 'too big' for the height of the extrusion. The error message is: 'Draft angle results in a self-intersecting body'.

Example 2 – a splined shaft

1 Open your A3SOL template file, UCS BASE, layer MODEL with the wport active.

2 Zoom centre about 0,0,40 at 200 magnification – all viewports.

3 Refer to Fig. 36.2 and create the layout from three circles of diameters 120, 20 and 16. The actual layout is your design.

4 Select the SUBTRACTION icon from the Modify II toolbar, pick the largest circle and:
prompt   No solids or regions selected

5 Select the REGION icon from the Draw toolbar and:
prompt   Select objects
respond   **pick all circles then right-click**
prompt   17 loops extracted
             17 Regions created.

6 Menu bar with **Modify–Boolean–Subtract** and:
prompt   Select solids or regions to subtract from
             Select objects
respond   **pick the largest circle then right-click**
prompt   Select solids or regions to subtract
             Select objects
respond   **pick the 16 smaller circles then right-click**
and   the region is created as fig. (a).

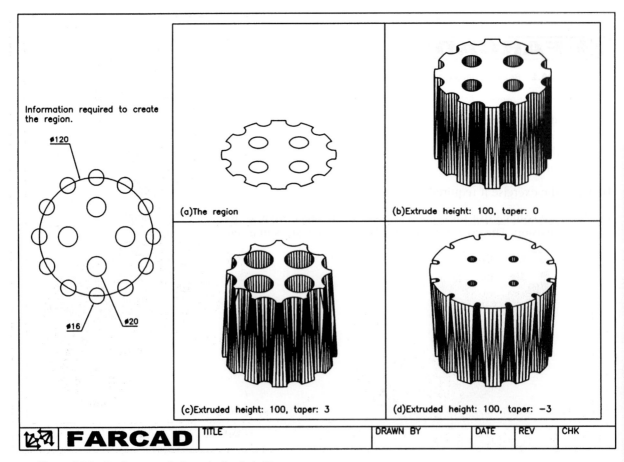

Information required to create the region.

ø120

ø16   ø20

(a)The region

(b)Extrude height: 100, taper: 0

(c)Extruded height: 100, taper: 3

(d)Extruded height: 100, taper: −3

FARCAD   TITLE   DRAWN BY   DATE   REV   CHK

Figure 36.2   Region example 2 – a 'splined shaft'.

7 At this stage save as R14MOD\REGEX for the next exercise.

8 With the lower left viewport active select the EXTRUDE icon from the Solids toolbar and:
   a) objects: pick the region
   b) height: 100
   c) taper angle: 0 – fig. (b).

9 Undo the extrusion effect, then use the EXTRUDE icon with the following entries:
   i) height: 100, taper angle: 3 – fig. (c)
   ii) height: 100, taper angle: −3 – fig. (d).

10 Hide, shade, save, etc.

Example 3 – a revolved component

1 Open drawing file R14MOD\REGEX saved from the previous exercise with UCS BASE
   – Fig. 36.3(a).

2 Menu bar with **Tools–UCS–Origin** and:
   prompt   Origin point<0,0,0>
   enter   **−100,−100,0 <R>**.

3 Zoom centre about 150, 0, 0 at 250 magnification.

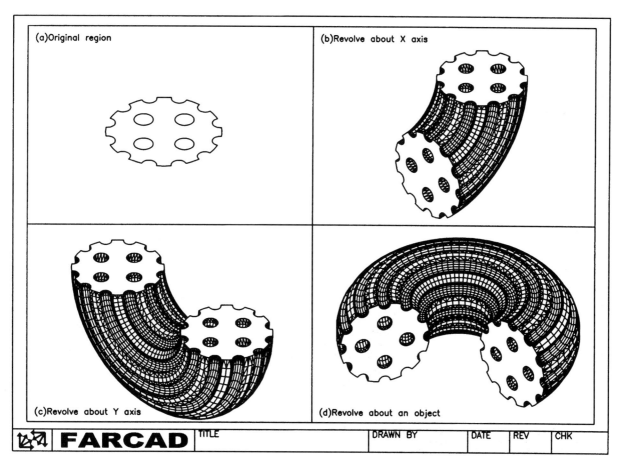

Figure 36.3 Region example 3 – a revolved 'component'.

4 Select the REVOLVE icon from the Solids toolbar and:
  prompt   `Select objects`
  respond **pick the region then right-click**
  prompt   `Axis of revolution - Object/X/Y...`
  enter   **X <R>** – the *X*-axis option
  prompt   `Angle of revolution<full circle>`
  enter   **−90 <R>**
  and   pan model to suit then hide – fig. (b).

5 Undo the hide and revolve effect to leave the original region.

6 Using the REVOLVE icon:
  a) pick the region then right-click
  b) enter Y as the axis if revolution
  c) enter 180 as the angle
  d) pan and hide – fig. (c)
  e) undo the hide and revolve effect.

7 Draw a line from: 0,0,0 to: @0,0,100.

8 Menu bar with **Modify–3D Operation–Rotate 3D** and:
  a) pick the region then right-click
  b) enter **X <R>** as the axis
  c) enter 100,100,0 as a point on the axis
  d) enter 90 as the rotation angle.

9 With the REVOLVE icon:
  a) pick the rotated region the right-click
  b) enter **O <R>** – object option
  c) pick lower end of vertical line
  d) enter 240 as the angle of revolution
  e) pan to suit then hide – fig. (d).

10 This completes the third exercise. Save if required.

Example 4

1 Open your A3SOL template file with UCS BASE, layer MODEL and the lower right viewport active. Refer to Fig. 36.4.

2 Draw three circles:
  a) centre: 50,0, radius: 50
  b) centre: 0,50, radius: 60
  c) centre: 75,75, radius: 75.

3 Make a new layer: BND, colour blue and current.

4 Zoom centre about 50,50,50 at 200 magnification.

5 Menu bar with **Draw–Boundary** and:
  prompt   Boundary Creation dialogue box – Fig. 36.5
  respond 1. pick Object Type: Region
             2. pick Pick Points<
  prompt   `Select internal point`
  respond **pick point indicated in Fig. 36.4**
  prompt   `Selecting everything...`
             `Selecting everything visible...`
             `Analyzing internal islands`

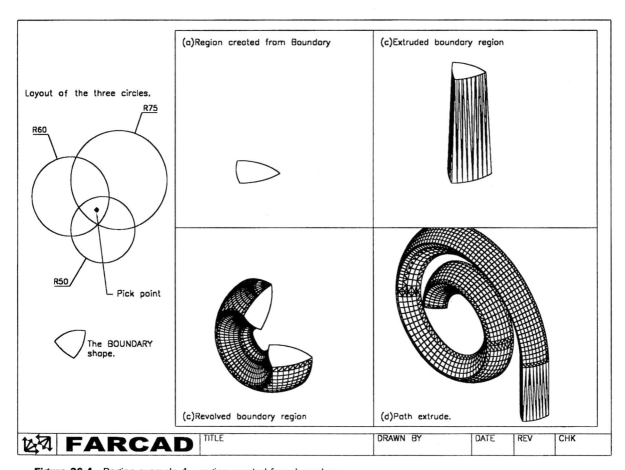

Figure 36.4 Region example 4 – region created from boundary.

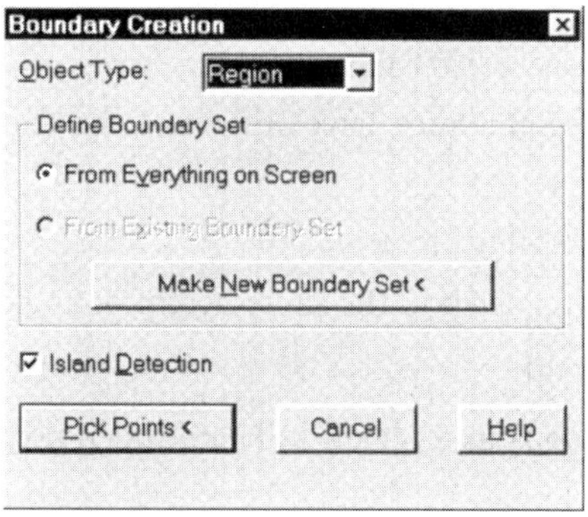

Figure 36.5 Boundary Creation dialogue box.

then	Select internal point
respond	**right-click**
prompt	1 loop extracted
	1 Region created
	BOUNDARY created 1 region
and	a blue region is displayed.

6 Erase the three circles to leave the blue region – fig. (a).

7 Using the EXTRUDE icon:
   a) pick the blue boundary region then right-click
   b) enter a height of 125
   c) enter a taper angle of 2
   d) hide the model – fig. (b).

8 Undo the hide and extrude effects to leave the blue region.

9 With the REVOLVE icon:
   a) select the blue region then right-click
   b) enter Y as the axis of revolution
   c) enter 270 as the angle of revolution
   d) hide – fig. (c).

10 Undo the hide and revolve effect.

11 Restore UCS FRONT, and with layer MODEL current draw a polyline:
from	0,0
to	@0,100
to	arc endpoint: @–200,0
to	arc endpoint: @120,0
to	arc endpoint: @–60,0 then right-click.

12 Restore UCS BASE and make layer BND current.

13 With the EXTRUDE icon:
   a) select the blue region
   b) enter **P <R>** for the path option
   c) pick the red polyline
   d) hide to give fig. (d).

14 The exercise is complete; save?

Summary

1 A region is created from closed shapes, e.g. polylines, arcs, circles, ellipses, etc.

2 Regions can be created using the BOUNDARY command.

3 Regions consist of loops and all regions must have an outer loop. There can be several inner loops.

4 Regions can be extruded and revolved.

5 All parts of a region are extruded/revolved to the same height or angle.

6 Regions can be extruded along a path.

7 *Note*: Release 14 does not allow loops in a region to be extruded/revolved to different heights/angles.

Assignment

This activity requires three regions to be created from polylines and then extruded to different heights.

Activity 20: Ratchet

The model consists of a base and two numbers:

a) base: made from a 60 radius circles and two lines. The lines are arrayed for nine items and then the trim command is used to complete the outline

b) number 1: simple

c) number 4: simple – use grid and snap to assist.

Notes

1 The arrayed base requires the edit polyline command to be used to 'convert' the outline into a single polyline.

2 The numbers can be made from polylines.

3 The numbers have to positioned on top on the ratchet base.

4 The extruded heights are:
base: 80
1: 50
4: 25.

Chapter 37

Moving solids

Solids can be moved and rotated with the normal 2D and 3D commands, but the ALIGN command is very useful. To demonstrate how it is used:

1 Open the A3SOL template file with UCS BASE, layer MODEL and the lower left viewport active.

2 Create the following three primitives:

Box	*Cylinder*	*Wedge*
corner: 0,0,0	centre: 150,150,0	corner: 150,0,0
cube	radius: 30	length: 80
length: 100	height: 75	width: 60
colour: red	colour: green	height: 60
		colour: blue.

3 Zoom centre about 50,50,75 at 250 magnification.

4 Copy the three primitives to another part of the screen.

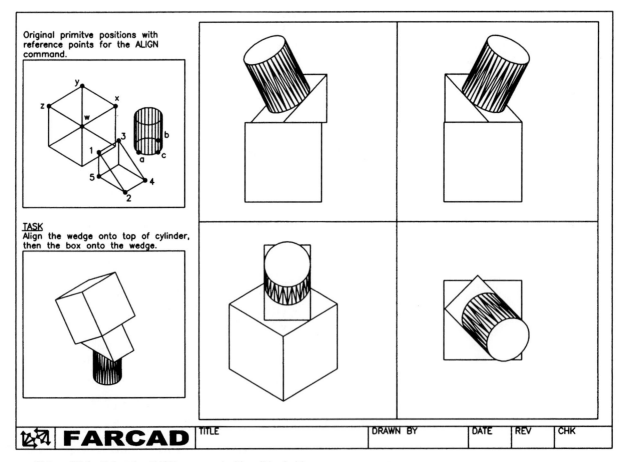

Figure 37.1 Using the ALIGN command with solid primitives.

5 The cylinder is to be aligned onto the wedge, and then the wedge and cylinder have to be aligned onto the top of the cube.

6 Refer to Fig. 37.1.

7 Menu bar with **Modify–3D Operation–Align** and:
   prompt   Select objects
   respond   **pick the green cylinder then right-click**
   prompt   Specify 1st source point
   respond   **QUADrant icon and pick base circle at pta**
   prompt   Specify 1st destination point
   respond   **MIDpoint icon and pick line 12**
   prompt   Specify 2nd source point
   respond   **QUADrant icon and pick base circle at ptb**
   prompt   Specify 2nd destination point
   respond   **MIDpoint icon and pick line 34**
   prompt   Specify 3rd source point
   respond   **QUADrant icon and pick base circle at ptc**
   prompt   Specify 3rd destination point
   respond   **MIDpoint icon and pick line 24**.

8 The cylinder will be aligned onto the wedge, with its base circle 'centred' on the sloped surface of the wedge.

9 Activate the ALIGN command again and:
   prompt   Select objects
   respond   **pick the blue wedge and green cylinder then right-click**
   prompt   Source and destination points
   and   *a*) 1st source point: pick ENDpoint pt4
               b) 1st destination point: pick MIDpoint line wx
               c) 2nd source point: pick ENDpoint pt2
               d) 2nd destination point: pick MIDpoint line zw
               e) 3rd source point: pick ENDpoint pt5
               f) 3rd destination point: pick MIDpoint line zy.

10 The wedge and cylinder will be aligned onto the top of the red cube – may need to pan in the viewports.

11 *Task*.
   Using the three copied primitives:
   a) align the long wedge surface onto the top of the cylinder
   b) align the cube onto any surface of the wedge.

12 The exercise is complete – hide, save?

Chapter 38

Enquiring into models

In this chapter we will create two new composites and then use the AutoCAD enquiry commands to determine the properties of the solids. We will also investigate how the material properties file is created.

Composite model 4 – a slip block

1 Open the A3SOL template file with UCS BASE, layer MODEL and the lower left viewport active.

2 Display the Inquiry toolbar and refer to Fig. 38.1.

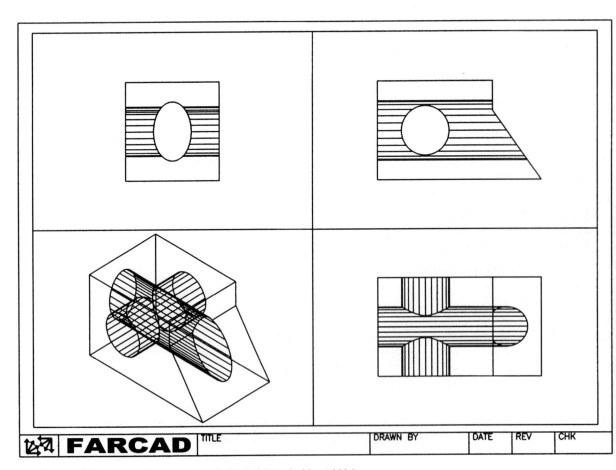

Figure 38.1 Composite model 4 – slip block (plotted without hide).

3 Create the following primitives:

Box	*Wedge*
corner: 0,0,0	corner: 120,0,0
length: 120	length: 50
width: 100	width: 100
height: 100	height: 70
colour: red	colour: blue.

4 Centre the model about 80,50,50 at 200 magnification.

5 Create two green cylindrical primitives with:
  a) centre: 60,0,50
      radius: 25
      centre of other end: @0,100
  b) elliptical option
      centre of ellipse: 0,50,50
      axis endpoint: @20,0
      other axis distance: @0,0,30
      centre of other end: @180,0.

6 *a*) union the red box and blue wedge
  b) subtract the two green cylinders from the composite
  c) note the 'curves of interpenetration'
  d) shade and note the colour effect, then regen.

7 At this stage save the composite as R14MOD\SLIPBL.

8 Select the AREA icon from the Inquiry toolbar and:

prompt	`<First point>/Object/...`
enter	**O <R>** – the object option
prompt	`Select objects`
respond	**pick the composite then right-click**
prompt	`Area = 98060.50, Perimeter = 0.00.`

9 This is the surface area of the composite in square units (mm?). A solid composite has no perimeter, hence the 0 value.

10 Select the MASS PROPERTY icon from the Inquiry toolbar and:

prompt	`Select objects`
respond	**pick the composite then right-click**
prompt	AutoCAD Text Window
with	details on the composite including:
	Mass = 996099.3322
	Volume = 996099.3322
respond	`<RETURN>`
prompt	`Write to a file?<N>`
enter	**Y <R>**
prompt	Create Mass and Area Properties File dialogue box
respond	1. check – Save in **r14mod** folder
	2. check – Save as type ***.mpr** – materials properties
	3. enter File name: **SLIPBL**
	4. pick Save – more on this file later.

11 The mass and volume values are the same, as AutoCAD R14 assumes a density of 1. Release 12 users should note that R14 does not support a materials library.

Composite model 5 – a casting block

1 A3SOL template file, UCS BASE, layer MODEL, lower left viewport active and:
   a) refer to Fig. 38.2
   b) zoom centre about 37,50,18 at 150 magnification.

2 Create a red box primitive with:
   corner: 0,0,0
   length: 75, width: 100, height: 36.

3 Create a red elliptical cylinder with:
   centre: 0,0,36
   axis endpoint: @20,0
   other axis distance: @0,35
   height: −18.

4 Rectangular array the elliptical cylinder:
   a) 2 rows and 2 columns
   b) row distance: 100, column distance: 75.

5 Subtract the four cylinders from the box – fig. (a).

6 Create another two red cylinders with:
   a) centre: 10,10
      radius: 4 and height: 18
   b) centre: 0,50,18
      radius: 10
      centre of other end: @75,0,0.

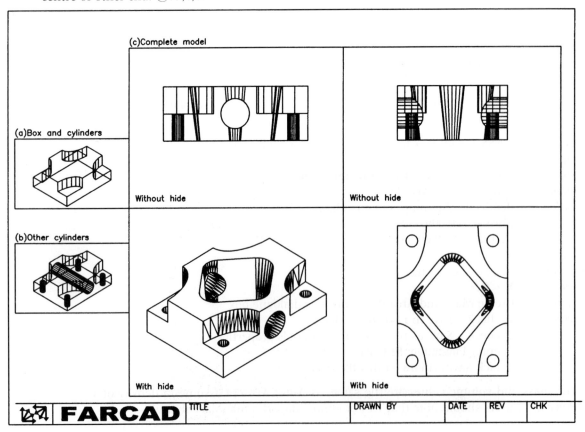

Figure 38.2 Composite model 5 – a casting block.

7 Rectangular array the smaller cylinder:
   a) for 2 rows and 2 columns
   b) row distance: 80, column distance: 55.

8 Subtract the five cylinders from the composite – fig. (b).

9 *a*) Draw a polyline:
      from: 0,0 to: 25,30 to: 0,60 to: −25,30 to: close
   b) Fillet the polyline with a radius of 6
   c) Move the polyline from: 0,0 by: @37.5,20
   d) Extrude the polyline for a height of 36 with −8 taper.

10 Complete the model by subtracting the extruded polyline from the box composite – fig. (c).

11 Save the model as R14MOD\CASTBL.

12 Menu bar with **Format–Units** and using the Units Control dialogue box:
   a) Units: Engineering
   b) Precision: 0′–0.00″.

13 Select the AREA icon from the Inquiry toolbar and:
   a) use the object option and pick the composite
   b) Area = 30824.09 square in (214.0562 square ft).

14 Select the MASS PROPERTIES icon, pick the composite and:

prompt	AutoCAD Text Window	
with	Mass	151409.88 lb
	Volume	151409.88 cu in
respond	**<RETURN>** to the prompt	
prompt	Write to a file?<N>	
enter	**Y <R>**	
prompt	Create Mass and Area Properties File dialogue box	
with	SLIPBL listed from previous exercise	
respond	1. enter file name: CASTBL – Fig. 38.3	
	2. pick Save.	

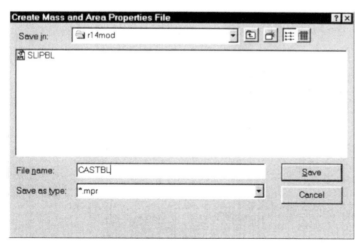

Figure 38.3 Create Mass and Area Properties File dialogue box.

15 *Investigate.*

The area and mass properties have been 'calculated' with the UCS BASE. There are two other UCS saved positions, FRONT and RIGHT. Using these saved UCS's, determine the area and mass/volume values with engineering units.

My values were:

UCS	Area(sq in)		Mass (lb)/Volume (cu in)
BASE	30824.0	9	151409.88
FRONT	30824.0	9	151410.30
RIGHT	30824.0	9	151409.73

These slight variations in mass/volume are due to the way in which R14 performs the calculations.

Mass properties file

When the mass properties icon is used with a solid model, the AutoCAD Text Window will display 'technical' information about the model including the mass and volume.

The user has the option of saving this information to a Mass Property file with the extension **.mpr**. As the mass property file is a 'text file' it can be opened in any 'text editor' type package. To demonstrate this:

1 Select the Windows Start icon from the Windows taskbar at the bottom of the screen.

2 Select **Programs–Accessories–Notepad** and:
   prompt Untitled Notepad screen
   respond 1. menu bar with **File–Open**
   2. double left click on r14mod
   3. alter file name to *.mpr
   4. pick SLIPBL
   5. pick Open.

3 The screen will display the saved material property file for the slip block composite – Fig. 38.4(a).

4 This file can then be printed if required.

5 Figure 38.4 also displays the material property file for the casting block.

6 Exit Notepad to return to AutoCAD.

7 The exercise is now complete.

Summary

1 Mass properties can be obtained from solid models.

2 The mass properties include mass, volume, centroid, radius of gyration, etc.

3 The mass properties can be saved to a .mpr text file.

4 Release 14 does not support a materials library and hence mass and volume are always the same, as density is assumed to be 1.

a)Material Properties File: SLIP BLOCK

---------------- SOLIDS ----------------

Mass: 996099.3322
Volume: 996099.3322
Bounding box: X: 0.0000 -- 170.0000
                 Y: 0.0000 -- 100.0000
                 Z: 0.0000 -- 100.0000
Centroid: X: 72.7670
                 Y: 50.0000
                 Z: 45.6038
Moments of inertia: X: 6601412624.6681
                 Y: 10308132625.0530
                 Z: 10617005572.1534
Products of inertia: XY: 3624157426.3442
                 YZ: 2271295936.3129
                 ZX: 3008236919.5127
Radii of gyration: X: 81.4080
                 Y: 101.7276
                 Z: 103.2404
Principal moments and X-Y-Z directions about centroid:
            I: 1942452034.2999 along [0.9506 0.0000 -0.3105]
            J: 2962157503.7769 along [0.0000 1.0000 0.0000]
            K: 2949494380.3586 along [0.3105 0.0000 0.9506]

b)Materials Properties File: CASTING BLOCK

---------------- SOLIDS ----------------

Mass: 151409.88 lb
Volume: 151409.88 cu in
Bounding box: X: 0.00 -- 75.00 in
                 Y: 0.00 -- 100.00 in
                 Z: 0.00 -- 36.00 in
Centroid: X: 37.50 in
                 Y: 50.00 in
                 Z: 15.30 in
Moments of inertia: X: 584593519.53 lb sq in
                 Y: 336682992.11 lb sq in
                 Z: 818548980.79 lb sq in
Products of inertia: XY: 283893527.78 lb sq in
                 YZ: 115810083.83 lb sq in
                 ZX: 86857562.86 lb sq in
Radii of gyration: X: 62.14 in
                 Y: 47.16 in
                 Z: 73.53 in
Principal moments (lb sq in) and X-Y-Z directions about centroid:
            I: 170636576.12 along [1.00 0.00 0.00]
            J: 88330609.83 along [0.00 1.00 0.00]
            K: 227104119.83 along [0.00 0.00 1.00]

Figure 38.4 Materials properties files from Notepad.

Chapter 39

Viewport-specific layers

Viewport-specific layers are exactly what their name suggests – they are layers 'tied' to a specific viewport. Viewport-specific layers have many uses including:
a) adding dimensions to a multi-view model
b) adding sections to a model.

The concept of viewport-specific layers was discussed during the creation of the first composite model, and in this chapter we will further investigate how to add dimensions to a composite model.

1　Open the drawing file R14MOD\SLIPBL created in a previous chapter.

2　Menu bar with **Format–Layer** and:
　　prompt　　Layer & Linetype Properties dialogue box
　　respond　　1. pick the Dim layer – highlighted in blue
　　　　　　　　2. pick New three times to make three new layers-Layer1, Layer2 and Layer3
　　　　　　　　3. rename these new layers as follows:
　　　　　　　　Layer1 – DIMTL
　　　　　　　　Layer2 – DIMTR
　　　　　　　　Layer3 – DIMBR
　　　　　　　　4. pick OK.

3　In model space:
　　a) make the top left viewport active and use the Layer Properties dialogue box to currently freeze layers Dimtr and Dimbr
　　b) with the top right viewport active, currently freeze layers Dimtl and Dimbr
　　c) with the lower right viewport active, currently freeze layers Dimtl and Dimtr
　　d) with the lower left viewport active, currently freeze the three new layers Dimtl, Dimtr and Dimbr.

4　Menu bar with **Dimension–Style** and:
　　prompt　　Dimension Styles dialogue box
　　with　　　3DSTD the current style
　　respond　　1. pick Geometry
　　　　　　　　2. alter Overall scale to 1.5 then pick OK
　　　　　　　　3. pick Save then OK.

5　With UCS BASE make the lower right viewport active and layer Dimbr current. Refer to Fig. 39.1.

6　Add the linear dimensions as shown:
　　a) one vertical
　　b) two horizontal using the baseline option.

7   With the top right viewport active, restore UCS FRONT and make layer Dimtr current.
    Add the given dimensions:
    a) two vertical baseline
    b) one angular
    c) one diameter – zoom needed?

8   With the top left viewport active, restore UCS RIGHT with layer Dimtl current and add
    the given dimensions.

9   Restore UCS BASE and save exercise if required.

Summary

1   Viewport-specific layers can be used to dimension a solid model.

2   A new layer should be made for each viewport that requires dimensions to be displayed.

3   Layers should be currently frozen in viewports where dimensions have not to be
    displayed.

4   Adding dimensions to a solid model is very dependent on the UCS.

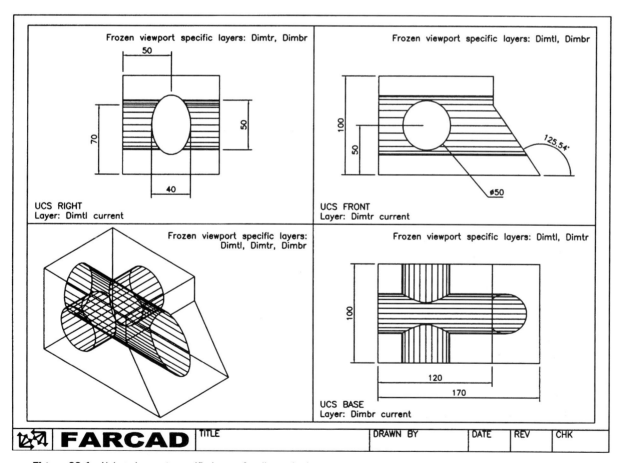

Figure 39.1   Using viewport-specific layers for dimensioning.

Assignment

To help you become proficient at making and using viewport-specific layers, the following activity should be attempted. The procedure is identical to the completed worked example.

Activity 21: Dimensioning the casting block

Use the R14MOD\CASTBL model created in a previous chapter and:
a) make three new viewport-specific layers for dimensions
b) add the dimensions displayed in the activity drawing.

Chapter 40

Slicing and sectioning solid models

Solid models can be sliced (cut) and sectioned relative to:
a) the three coordinate planes
b) three points defined by the user
c) the viewing plane of the current viewpoint
d) user-defined objects.

The two commands are very similar in operation and when used:
a) the slice command results in a new composite model. This model retains the layer and colour properties of the original solid
b) the section command adds a 2D region to the model. The region is displayed on the current layer but **no hatching is added to the region**. This is the same as Release 13, but differs from the section command in Release 12 where hatching is added to the region.

The two commands will be demonstrated using previously created composite models.

Slice example 1 – using the three coordinate planes

1 Open the composite model R14MOD\SLIPBL with UCS BASE, layer MODEL and the lower left viewport active.

2 Refer to Fig. 40.1 which only displays the 3D viewport for the exercise.

3 Select the SLICE icon from the Solids toolbar and:
  prompt   Select objects
  respond   **pick the composite then right-click**
  prompt   Slicing plane by Object/Zaxis/View/XY...
  enter   **XY <R>** – the XY plane option
  prompt   Point on XY plane<0,0,0>
  enter   **0,0,50 <R>** – why these coordinates?
  prompt   Both sides/<Point on desired side of plane>
  enter   **@0,0,–10 <R>**, i.e. keep part 'below' slicing plane
  and   a new model is created – fig. (a).

4 Undo the slice effect.

5 Menu bar with **Draw–Solids–Slice** and:
  a) pick the composite then right-click
  b) enter YZ <R> as the slicing plane
  c) enter 50,0,0 <R> as a point on the plane
  d) enter @–10,0,0 <R> as a point on side to 'keep'
  e) the new composite is created as fig. (b).

6 Undo the slice effect.

7 At the command line enter **SLICE <R>** and:
  a) pick the composite then right-click
  b) enter ZX as the slicing plane
  c) enter 0,50,0 as a point on the plane
  d) enter @0,10,0 as a point on desired side of plane
  e) new composite as fig. (c).

8 Undo the slice effect, activate the slice command and:
  a) pick the composite then right-click
  b) select the YZ plane
  c) enter 40,0,0 as a point on the plane
  d) enter **B <R>** for both sides
  e) model is now 'sliced' into two new composites.

9 With the MOVE icon:
  a) pick the blue wedge then right-click
  b) base point: enter 170,0
  c) second point of displacement: enter @50,0
  d) the two new composites are separated – fig. (d).

10 This completes the first exercise which does not need to be saved.

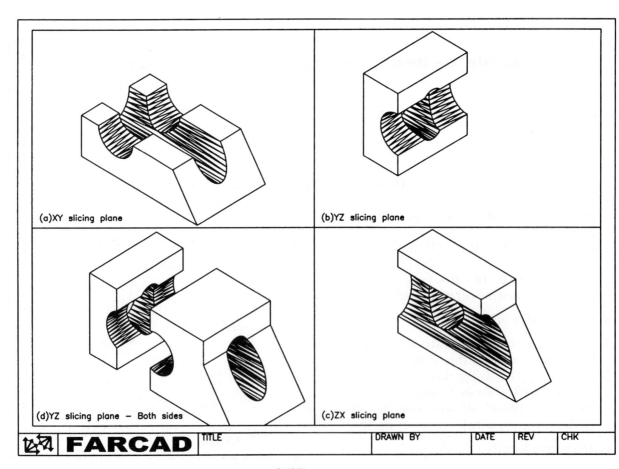

Figure 40.1   Slice example 1 – composite model SLIPBL.

Slice example 2 – using user-defined slicing planes

1 Open the drawing file R14MOD\MACHSUPP – the first composite solid created, with USC BASE, layer MODEL and the lower left viewport active. Refer to Fig. 40.2 which displays the 3D viewport for the exercise.

2 Activate the SLICE command and:

prompt	Select objects
respond	**pick the composite then right-click**
prompt	Slicing plane by Object...
respond	**right-click** – accepting the 3 point default
prompt	1st point on plane and: **ENDpoint and pick pt1**
prompt	2nd point on plane and: **ENDpoint and pick pt2**
prompt	3rd point on plane and: **ENDpoint and pick any pt3 on cylinder**
prompt	Both sides/<Point on desired side of plane>
enter	**@0,0,–10** <R> – why this entry?

3 A new composite will be created as fig(a). This has been 'sliced' through an inclined plane.

4 Hide and shade the model.

5 Undo the shade, hide and slice commands to leave the original composite – or re-open MACHSUPP.

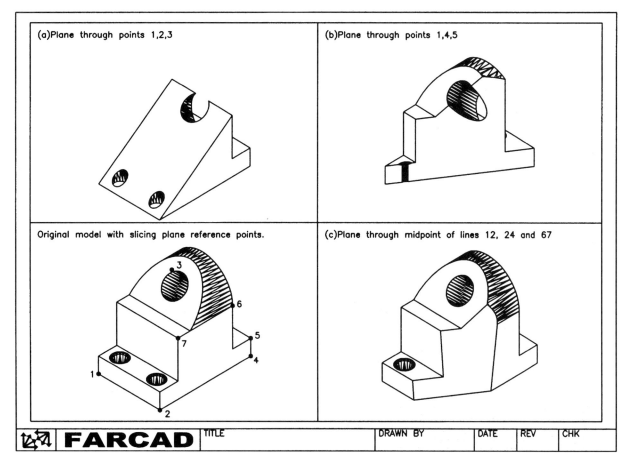

Figure 40.2 Slice example 2 – using three-defined points.

6 With the SLICE command:
   a) pick the composite then right-click
   b) enter <R> to activate the 3 point option
   c) 1st point on plane: ENDpoint and pick pt1
   d) 2nd point on plane: ENDpoint and pick pt4
   e) 3rd point on plane: ENDpoint and pick pt5
   f) point on plane: enter @0,10,0
   g) hide the new composite – fig. (b).

7 This new composite has been sliced through a vertical plane from corner to corner.

8 Undo the hide and slice effects.

9 Use the SLICE command on the composite, activate the three point option and:
   a) 1st point on plane: MIDpoint of line 12
   b) 2nd point on plane: MIDpoint of line 24
   c) 3rd point on plane: MIDpoint of line 67 – take care with this!
   d) point on plane: enter: @**−10,0,0**
   e) new composite displayed as fig. (c)
   f) hide and shade.

10 Exercise is now complete.

Slice example 3 – view and object options

1 Open the drawing file R14MOD\FLPIPE of the flange-pipe composite from Chapter 33, UCS BASE and layer MODEL.

2 Refer to Fig. 40.3 and set the following 3D viewpoints:
   viewport   *viewpoint*
   top left   NW Isometric
   top right   NE Isometric
   bottom right   SW Isometric
   bottom left   SE Isometric – should be set to this.

3 With the top left viewport active, select the SLICE icon and:
   prompt   Select objects
   respond **pick the composite then right-click**
   prompt   Slicing plane by Object/...
   enter   **V <R>** – the view option
   prompt   Point on view plane<0,0,0>
   enter   **0,50,0 <R>**
   prompt   Both sides/<Point on desired side of plane>
   enter   **0,−10,0 <R>**
   and   new composite displayed as fig. (a)
   then   hide the model.

4 *Note*: the view option assumes a 'line of sight' which is perpendicular to the viewpoint plane.

5 Undo the hide and slice effects to restore the original model.

6 With the top right viewport active, SLICE and:
   a) pick the composite then right-click
   b) enter V <R> for the view option
   c) enter 0,50,0 as a point on the view plane
   d) enter 0,−10,0 as a point on the desired side on the plane
   e) new composite created
   f) hide the model – fig. (b).

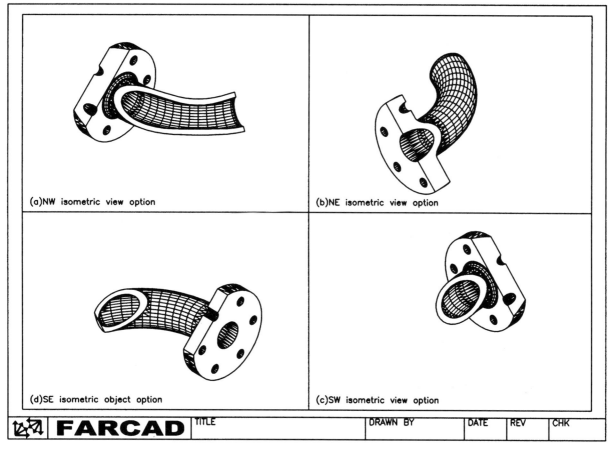

(a)NW isometric view option

(b)NE isometric view option

(d)SE isometric object option

(c)SW isometric view option

FARCAD | TITLE | DRAWN BY | DATE | REV | CHK

Figure 40.3 Slice example 3 – using view and object options.

7 Undo the hide and slice effects.

8 Make the lower right viewport active and slice with:
   a) the composite
   b) the view option
   c) 0,–50,0 as a point on the plane
   d) 0,10,0 as a point on the desired side of the plane
   e) new composite as fig. (c)
   f) hide then undo both the hide and slice effects.

9 Finally with the lower left viewport active:
   a) draw a circle with centre: 0,–100 and radius: 100
   b) rotate 3D this circle:
      i)   about the X axis
      ii) with the circle centre as a point on the axis
      iii) for a reference angle of 45 degrees
   c) with the SLICE command:
      i)   pick the composite then right click
      ii) enter **O <R>** for the object option
      iii) pick the circle as the object
      iv) enter 0,10,0 as a point on desired side
      v)   composite as fig. (d).

10 The slice examples are now complete.

Section example – three point option

The section command is very similar in operation to the slice command, and only one example will be demonstrated.

1 Open the drawing file R14MOD\SLIPBL from Chapter 38, UCS BASE, layer MODEL and the lower left viewport active. Refer to Fig. 40.4.

2 Make layer SECT current – or make a new section layer.

3 Select the SECTION icon from the Solids toolbar and:

    prompt   `Select objects`
    respond   **pick the composite then right-click**
    prompt   `Section plane by Object/...`
    respond   **right-click** – to activate the three point option
    prompt   `1st point on plane` and pick ENDpoint of pt1
    prompt   `2nd point on plane` and pick ENDpoint of pt2
    prompt   `3rd point on plane` and pick ENDpoint of pt3.

4 A region is added to the model – the section plane.

5 *Note.*
    a) Hatching is not automatically added to the region in R14. This is also true with the section command in R13, but not with R12. Release 12 automatically adds hatching to the region.
    b) Hatching must be added to the section region by the user.

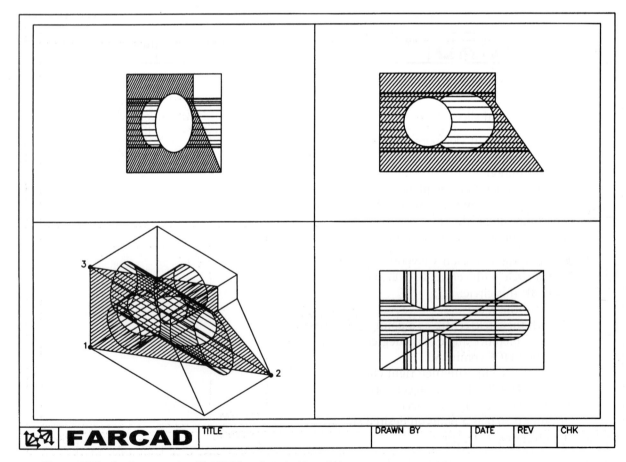

Figure 40.4 Section example with SLIPBL – plotted without hide.

c) The section region is displayed in all viewports, and the viewport specific layer concept is used to 'currently freeze' the section in specific viewports.

d) If hatching is to be added to the section region (as it should?), remember to alter the UCS position.

e) My section region in Fig. 40.4 has hatching added using the following variable values – HPNAME: ANSI31, HPSCALE: 1 and HPANG: 0.

f) I have plotted the viewport configuration without hide – why?

Using the slice and section commands

If a 'true section' has to be obtained from a solid composite, then the slice and section commands can both be used. To demonstrate this concept:

1 Open drawing file R14MOD\CASTBL from Chapter 38 with UCS BASE, layer MODEL and the lower left viewport active.

2 The original model is displayed in Fig. 40.5(a).

3 With the SLICE command:
   a) pick the composite then right-click
   b) use the three points on slicing plane option and pick the ENDpoints of points 1,2 and 3
   c) enter 0,100,0 as the point on the desired side on the plane
   d) a new composite will be displayed.

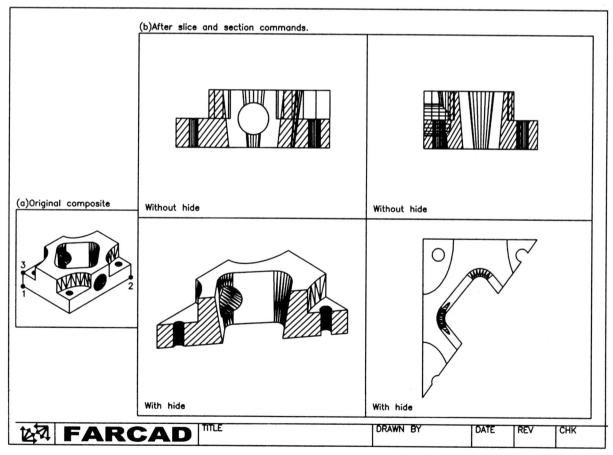

Figure 40.5 Using the SLICE and SECTION commands with CASTBL.

4 With layer SECT current, use the SECTION command and:
   a) pick the new composite then right-click
   b) use the three point options with the same selection as step 3
   c) region added to composite.

5 Hatch the region with correct UCS setting.

6 Figure 40.5 displays the result of the SLICE and SECTION commands with hide.

Summary

1 The SLICE command produces a new composite on the current layer.

2 The SECTION command adds a region to the section plane of the model, but does not add any hatching to this region. Hatching must be added by the user and is UCS dependent.

3 Both commands are very similar with the following options:
   a) the *XY*, *YZ* and *ZX* slicing/sectioning planes
   b) defining any three points on the slice/section plane
   c) relative to an object
   d) relative to the viewing plane in the current viewpoint.

4 The SLICE command requires:
   a) a point on the slicing plane
   b) a point on the desired side of the plane which is to 'be kept'.

5 The SECTION command only requires a point on the sectioning plane.

6 The position of the UCS can vary the *XY*, *YZ* and *ZX* planes.

7 The two commands can be activated:
   a) by icons from the Solids toolbar
   b) from the menu bar with Draw–Solids
   c) by entering SLICE or SECTION from the keyboard.

Assignment

This assignment requires the model from activity 19 to be used.

Activity 22: MODCOMP composite

Use the fillet/chamfer model from activity 19 to create a new composite using the plane information 1–2–3 or 1–4–5 as shown. The model has to be created from the section and slice commands, and I have only shown the 3D viewport.

Note. I have displayed the two composites on the one drawing using viewport specific layers. You may not be able to obtain this effect. It requires the original model to be copied and changed onto new viewport specific layers.

Chapter 41

The PROFILE command

When the HIDE command is used with a solid composite, the model is displayed with hidden line removal – as expected. From an 'engineering viewpoint' this is not what is wanted, as the front, top and end views should have lines which represent hidden detail. Release 14 allows this hidden detail to be 'added' to a solid model with the PROFILE command and when used, new layers are automatically added to the existing drawing.

A profile is defined as: an image which displays the edges of solid surfaces in the current view.

The command will be demonstrated with a previously created model, and then with a new detailed solid composite.

Example 1 – slip block

1 Open drawing R14MOD\SLIPBL from Chapter 38. This model has been used in several other chapters.

2 In paper space, use the LIST command and pick the top right viewport border. The AutoCAD Text Window will display information about the viewport including the **HANDLE** number. Take a note of this handle number, then repeat the LIST command selecting the other three viewport borders. In each case note the handle as it will be referred to in the exercise. My handle numbers for the four viewports were:
top right: 8A   top left: 8C
lower right: 90   lower left: 8E

3 *Note*.
 a) all objects created in AutoCAD are given a handle number and this number increase every time a new object is added to the drawing.
 b) the handle numbers are in hexadecimal format – don't worry about this if you have never heard of it. It is not important at this level of AutoCAD.
 c) the handles are for assisting with the AutoCAD database.
 d) your four viewport handle numbers may differ from mine. This is perfectly normal. Simply note them.

4 After the viewport border handle numbers have been noted, return to paper space with UCS BASE.

5 Make layer 0 current and the top left viewport active.

6 Refer to Fig. 41.1.

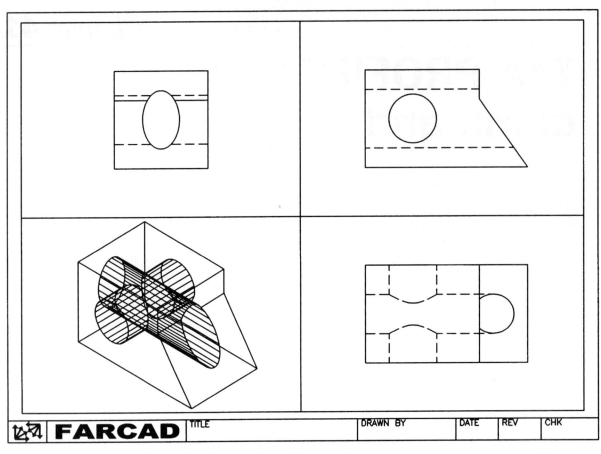

Figure 41.1 Using the PROFILE command with SLIPBL.

7 Select the SETUP PROFILE icon from the Solids toolbar and:

prompt	Select objects
respond	**pick the composite then right-click**
prompt	Display hidden profile lines on a separate layer?<Y>
enter	**Y <R>**
prompt	Project profile lines onto a plane?<Y>
enter	**Y <R>**
prompt	Delete tangential edges?<Y>
enter	**Y <R>**
and	the model in the top left viewport will be displayed with black lines over the model lines.

8 Menu bar with **Format–Layer** and:

prompt	Layer Control dialogue box
with	two new layers:
	a) **Ph-8c** for hidden profile lines – Hidden linetype?
	b) **Pv-8c** for visible profile lines – Continuous
respond	1. check the linetype for layer Ph-8c
	2. if it is not Hidden, then LOAD the HIDDEN linetype and set it to layer Ph-8c
	3. currently freeze layer MODEL
	4. pick OK
note	these two new layers will have the same handle number as top left viewport, in my case 8C.

9   The top left viewport will only display the profile visible and hidden lines for the composite as layer MODEL has been currently frozen in the active viewport. Do you understand why the linetype for layer Ph-8c was set to HIDDEN?

10 Make the top right viewport active and from the menu bar select **Draw–Solids–Setup–Profile** and:

   prompt   Select objects
   respond **pick the composite then right-click**
   prompt   Display hidden profile lines... and right-click
   prompt   Project profile lines... and right-click
   prompt   Delete tangential edges... and right-click.

11 The model will be displayed with black visible and hidden lines.

12 Menu bar with **Format–Layers** and:

   prompt Layer Control dialogue box
   with   Two new layers:
             a) Ph-8a with HIDDEN linetype
             b) Pv-8a with CONTINUOUS linetype
   respond 1. currently freeze layer MODEL
             2. pick OK
   note   remember that Ph-8a and Pv-8a are the hidden and visible layers for my viewport handles. You may have different Ph and Pv handle numbers.

13 The top right viewport will display visible and hidden lines for the model.

14 Make the lower right viewport active, still with layer 0 current.

15 At the command line enter **SOLPROF <R>** and:
   a) pick the model then right-click
   b) enter Y <R> to the display hidden profile lines prompt
   c) enter Y <R> to the project profile lines prompt
   d) enter Y <R> to the delete tangential edges prompt
   e) layer control dialogue box and:
       i)   two new layers: Ph-90 and Pv-90
       ii) currently freeze layer MODEL
       iii) pick OK.

16 The viewport will display the visible and hidden detail lines for the model in the viewport.

17 At this stage the layout should resemble Fig. 41.1 and can now be saved if required. This exercise is now complete.

18 *Explanation*.
   The PROFILE command is **viewport specific** and when activated:
   a) two viewport specific layers are automatically created, these being **Ph-??** and **Pv-??**
   b) the Ph layer is for hidden detail
   c) the Pv layer is for visible detail
   d) the ?? with the layer name is the current viewport handle number and is not controlled by the user
   e) the Pv linetype is continuous
   f) the Ph linetype is hidden, but **must be loaded**
   g) the command must be used in each viewport in which profile detail has to be extracted
   h) the command is generally used in viewports which display top, front and side views of the model
   i) the command is generally not used in a viewport which displays a 3D view of the model
   j) the hidden and visible detail added are blocks, i.e. there is a hidden detail block and a visible detail block. These blocks can be exploded if required.

Example 2 – a desk tidy as a detailed drawing

In this example a new composite will be created and saved for future recall. The model will be used to extract a section view and a view containing hidden detail, so open your A3SOL template file with UCS BASE, layer MODEL, lower left viewport active and refer to Fig. 41.2.

Altering the viewports

1 In paper space select the STRETCH icon from the Modify toolbar and:

prompt	Select objects
enter	**C <R>** – the crossing option
prompt	First corner and enter: **180,130 <R>**
prompt	Other corner and enter: **240,180 <R>**
prompt	4 found, Select objects
respond	**right-click**
prompt	Base point or displacement and enter: **210,155 <R>**
prompt	Second point of displacement and enter: **@–25,25 <R>**.

2 The viewport configuration will be altered.

3 Return to model space with UCS BASE.

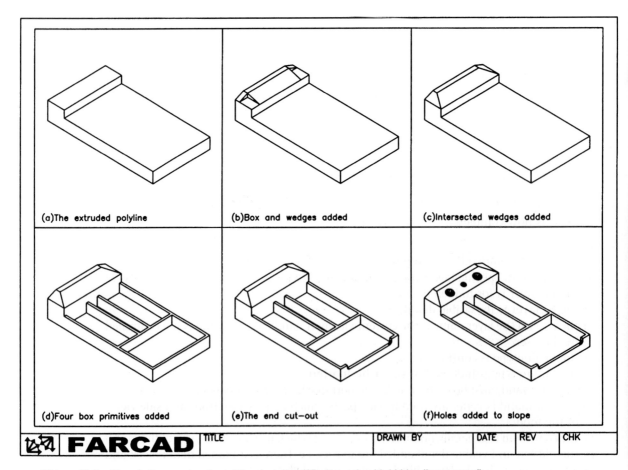

(a)The extruded polyline

(b)Box and wedges added

(c)Intersected wedges added

(d)Four box primitives added

(e)The end cut-out

(f)Holes added to slope

FARCAD | TITLE | DRAWN BY | DATE | REV | CHK

Figure 41.2 Steps in the construction of the desk tidy (3D view only with hidden line removal).

The basic shape

1 With the lower left viewport active restore UCS FRONT.

2 Draw a polyline using the following entries:
   From 0,0   *to* @156,0   *to* @0,15   *to* @–132,0
   to   @0,10   *to* @–24,0   *to* close.

3 With the EXTRUDE icon from the Solids toolbar, extrude the red polyline with:
   a) height: **–85**
   b) taper: **0**.

4 The extruded polyline will be displayed in 3D as fig. (a).

5 Restore UCS BASE.

6 In each viewport, zoom centre about 78,42,8 at **1XP** – yes 1XP!

The top

1 Create a box primitive with:
   a) corner: 0,12,25
   b) length: 6; width: 61; height: 5.

2 Create three wedges using the following information:
   a) corner: 6,12,25; length: 18; width: 61; height: 5
   b) corner: 6,12,25; length: 12; width: –6; height: 5
   c) corner: 6,73,25; length: 12; width: 6; height: 5.

3 With the ROTATE icon, rotate the following wedges:
   i) wedge (b) about the point 6,12,25 by **–90**
   ii) wedge (c) about the point 6,73,25 by **90**.

4 Union the box and the three wedges with the red extrusion.

5 The model at this stage resembles fig. (b).

6 Create another two wedges with:
   a) corner: 6,0,25; length: 18; width: 12; height: 5
   b) corner: 24,12,25; length: 12; width: –18; height: 5.

7 Rotate the second wedge (b) about the point 24,12,25 by **–90**.

8 Menu bar with **Modify–Boolean–Intersect** and:
   prompt   Select objects
   respond **pick the two wedges then right-click**.

9 Menu bar with **Modify–3D Operation–Mirror 3D** and:
   prompt   Select objects
   respond **pick the intersected wedges then right-click**
   prompt   Plane by Object/...
   enter   **ZX <R>** – the ZX plane option
   prompt   Point on ZX plane
   enter   **0,42.5,25 <R>**
   prompt   Delete old objects?<N> and enter: **N <R>**.

10 Union the two sets of intersected wedges with the composite.

11 The model should be displayed as fig. (c).

The compartments

The desk tidy compartments will be created from boxes subtracted from the composite.

1 With lower left viewport active and UCS BASE, create the following four box primitives:

	box 1	box 2	box 3	box4
corner	153,3,3	100,3,3	100,36,3	100,52,3
length	−50	−76	−76	−76
width	79	30	13	30
height	20	20	20	20
colour	magenta	blue	green	blue.

2 Subtract the four boxes from the composite – fig. (d).

3 Try a hide and shade (impressive?) the regen.

The end cut-out

1 Lower left viewport with UCS BASE?

2 Set a new UCS position with the three point option using:
   a) origin: 156,0,0
   b) x-axis position: 156,85,0
   c) y-axis position: 156,0,15
   d) save UCS position as NEWEND.

3 Draw a polyline with the following keyboard entries:
From	10,15
To	@0,−3
To	arc option with arc endpoint: @3,−3
To	line option with line endpoint: @59,0
To	arc option with arc endpoint: @3,3
To	line option with line endpoint: @0,3
To	close.

4 Extrude the polyline for a height of −3 with 0 taper.

5 Subtract the extruded polyline from the composite – fig. (e).

6 Restore UCS BASE.

The holes on the slope

1 In paper space zoom in on the 3D sloped area then model space.

2 Set a new UCS position with the three point option using:
   a) origin: 24,42.5,25
   b) x-axis position: 24,85,25
   c) y-axis position: 6,42.5,30
   d) save UCS position as SLOPE.

3 Create two primitives:

	cylinder	cone
centre	0,9.34077,0	0,9.34077,−12
radius	3	3
height	−12	−3
colour	magenta	magenta.

4 Union the cone and cylinder and then subtract then from the composite.

5 Create another two magenta cylinders:
   a) centre: 20,9.34077,0; radius: 5; height: −20
   b) centre: −20,9.34077,0; radius: 5; height: −20.

6 Subtract these two magenta cylinders from the composite – fig. (f).

7 In paper space zoom previous then return to model space.

8 Restore UCS BASE.

The complete model

1 The model is now complete and your screen display should be similar to Fig. 41.3.

2 Save the model at this stage as R14MOD\DESKTIDY.

3 Try the hide and shade commands in each viewport but remember to REGENALL.

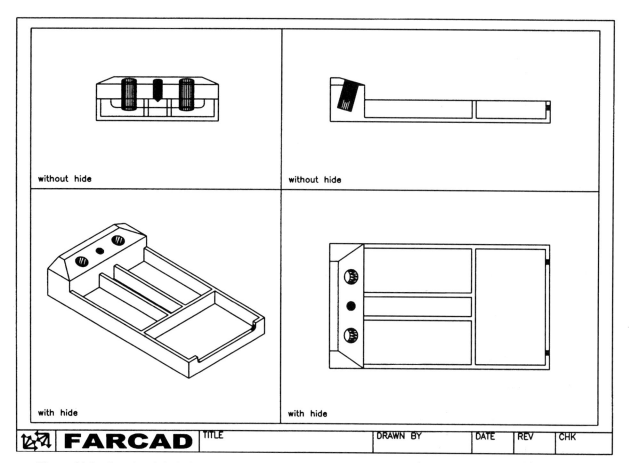

Figure 41.3 Completed desk tidy composite.

Extracting the profiles

1 Refer to Fig. 41.4.

2 With the top left viewport active, select the SETUP PROFILE icon from the Solids toolbar and:

prompt	Select objects
respond	**pick the composite then right-click**
prompt	Display hidden profile lines... and enter: **Y <R>**
prompt	Project profile line... and enter: **Y <R>**
prompt	Delete tangential edges and enter: **Y <R>**.

3 The model will be displayed with black lines.

4 Using the layer control dialogue box:
   a) note two new layers Ph-8c and Pv-8c (or similar with your viewport handle number
   b) load the linetype HIDDEN if required and set layer Ph-8c with this linetype
   c) current freeze layer MODEL in this viewport.

5 With the top right viewport active repeat the PROFILE command and pick the composite.

6 Using the layer control dialogue box:
   a) note two additional layers Ph-8a and Pv-8a (or similar)
   b) layer Ph-8a should have HIDDEN linetype as it has already been loaded
   c) currently freeze layer MODEL in this viewport.

7 Optimize the LTSCALE variable to display the hidden profile detail to your requirement.

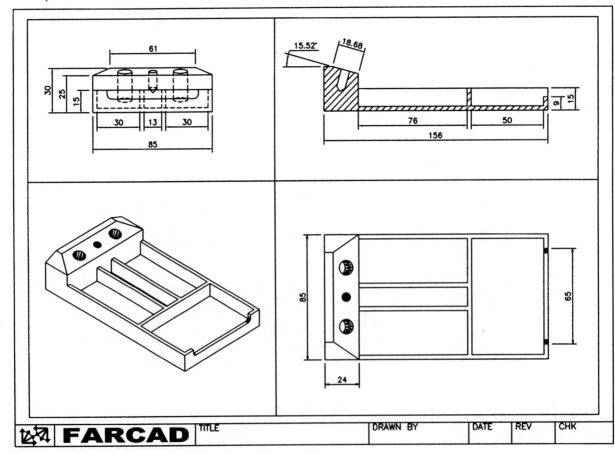

Figure 41.4   Extracted details from the desk tidy.

Extracting the section

1 With the lower right viewport active, ensure UCS BASE and make layer SECT current.

2 Using the SECTION icon from the Solids toolbar:
   a) pick the composite then right-click
   b) enter ZX as the section plane
   c) enter 0,42,5,0 as a point on the plane
   d) a region will be displayed in all viewports.

3 *a*) Make the top right viewport active
   b) currently freeze layer SECT in the other three viewports.

4 With an appropriate UCS setting, add hatching to the region using: pattern name: ANSI31; scale: 1; angle: 0.

Task

1 The profile extraction in the top right viewport is 'not quite correct' for the view displayed, so:
   a) erase the hidden detail block
   b) erase the visible horizontal line in the hatch area – remember that the visible detail is a block
   c) should there be another line displayed at the hole?

2 Using viewport specific layers, add the dimensions as Fig. 41.4.

3 Interrogate the model for:
   a) area: 47445.30
   b) mass: 106311.43.

4 Freeze the viewport layer and plot.

5 The detail exercise is now complete.

Profile explanation

When the profile command is used with a solid model, three prompts are displayed and it is usual to enter Y to these prompts.

a) *Display hidden profile lines on separate layers*. This creates two blocks, one for visible lines and one for hidden lines. Two new viewport specific layers are created for this block information Ph-?? and Pv-??. The actual ?? number is the handle of the current viewport, i.e. it is unique. The Pv (visible detail) has a continuous linetype, while the Ph (hidden detail) has a hidden linetype. The hidden linetype must be loaded before it can be 'assigned' to the Ph layer.

b) *Project profile lines onto a plane*. The profile detail is displayed as 2D objects and are projected onto a plane perpendicular to the viewing direction and passing through the UCS origin.

c) *Delete tangential edges*. A tangential edge is an imaginary edge where two faces meet at a tangent. Tangential edges are not shown for most drawing applications.

Chapter 42

Blocks and solid modelling

Blocks can be created and inserted with solid models as with any other 2D or 3D object. In this chapter we will create two interesting (I hope) solid model assembly drawings from blocks and then investigate solid model external references. We will also investigate the 'interference' between solids.

Example 1 – a desk tray assembly

1 Open your A3SOL template file with UCS BASE, layer MODEL and the lower left viewport active. Refer to Fig. 42.1.

2 Zoom centre about 55,40,20 at 0.75XP in all viewports.

3 Set the ISOLINES system variable to 12.

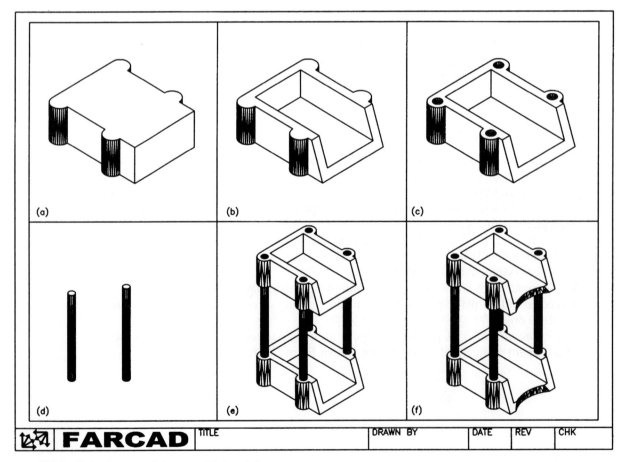

Figure 42.1 Creation of the assembled model.

The Tray

1 Make a new layer TRAY, colour blue and current.

2 Create two primitives from:
   Box   *Cylinder*
   corner: 0,0,0   centre: 10,0,0
   length: 110   radius: 10
   width: 80   height: 40
   height: 40.

3 Rectangular array the cylinder:
   a) for 2 rows and 2 columns
   b) row distance: 80 and column distance: 70.

4 Union the box and the four cylinders – fig. (a).

5 Create another two primitives from:
   Box   *Wedge*
   corner: 10,10,10   corner: 110,0,40
   length: 110   length: –15
   width: 60   width: 80
   height: 40   height: –40.

6 Subtract the box and wedge from the composite – fig. (b).

7 Create a cylinder with centre: 10,0,0; radius: 5 and height: 10.

8 Menu bar with **Modify–3D Operation–3D Array** and:
   prompt   Select objects
   enter   **L <R><R>** – two returns for the cylinder (last object)
   prompt   Rectangular or polar and enter: **R <R>**
   prompt   Number of rows and enter: **2 <R>**
   prompt   Number of columns and enter: **2 <R>**
   prompt   Number of levels and enter: **2 <R>**
   prompt   Row distance and enter: **80 <R>**
   prompt   Column distance and enter: **70 <R>**
   prompt   Level distance and enter: **30 <R>**.

9 Subtract the eight cylinders from the composite – fig. (c). A paper space zoom may be needed to help with this?

Making the tray block

At the command line enter **BLOCK <R>** and:
prompt   Block name and enter: **TRAY <R>**
prompt   Insertion base point and enter: **0,0,0 <R>**
prompt   Select objects
respond   **pick the composite then right-click.**

The support

1 Make a new layer SUPPORT, colour red and current.

2 Create two cylinders from:
   Cylinder 1   *Cylinder 2*
   centre: 0,0,0   centre: 50,50,0
   radius: 5   radius: 5
   height: 130   height: 140.

3 Change the colour of the 140 high cylinder to green.

4 The two cylinders are displayed as fig. (d).

5 Make two blocks of these cylinders with:
   a) Name: SUP1
      Insertion base point: 0,0,0
      Objects: pick the red cylinder
   b) Name: SUP2
      Insertion base point: 50,50,0
      Objects: pick the green cylinder.

Inserting the blocks

1 Zoom centre about 55,40,90 at 0.5XP in all viewports.

2 Make layer TRAY current.

3 Menu bar with **Insert–Block** and:
   prompt   Insert dialogue box
   respond   **pick Block**
   prompt   Defined blocks dialogue box
   respond   **pick TRAY then OK**
   prompt   Insert dialogue box with TRAY as block name
   respond   **pick OK**
   prompt   Insertion point and enter: **0,0,0 <R>**
   prompt   X scale factor and enter: **1 <R>**
   prompt   Y scale factor and enter: **1 <R>**
   prompt   Rotation angle and enter: **0 <R>**.

4 Repeat the INSERT command and insert block TRAY:
   a) at the point 0,0,150
   b) full size (i.e. X=Y=1) with 0 rotation.

5 At the command line enter **INSERT <R>** and:
   prompt   Block name and enter: **SUP2 <R>**
   prompt   Insertion point and enter: **10,0,30 <R>**
   prompt   X scale factor and enter: **1 <R>**
   prompt   Y scale factor and enter: **1 <R>**
   prompt   Rotation angle and enter: **0 <R>**.

6 Menu bar with **Modify–Boolean–Union** and:
   prompt   Select objects
   respond   **pick the three inserted blocks then right-click**
   prompt   At least 2 solids or coplaner regions must be selected.

7 What does this prompt mean? The union operation has not been successful with the three inserted blocks. Blocks must be exploded before they can be used with Boolean operations.

8 Using the EXPLODE icon from the Modify toolbar, select the three inserted blocks.

Checking for interference

1 Make layer 0 current.

2 Select the INTERFERE icon from the Solids toolbar and:
   prompt   Select the first set of solids
   respond   **pick the top blue tray then right-click**
   prompt   Select the second set of solids
   respond   **pick the green support then right-click**
   prompt   Comparing 1 solid against 1 solid
              Interfering solids (first set) : 1
              (second set)   : 1
              Interfering pairs   : 1
              Create interference solids><N>
   enter   **Y <R>**.

3 The model will be displayed with a black cylinder (layer 0) where interference occurs between the tray and the support leg. This interference is due to the support leg (SUP2) being too long, or the TRAY being inserted into the drawing at the wrong insertion point. We deliberately created the support leg (SUP2) with a height of 140 to obtain interference.

4 Erase the green support leg and the black interference cylinder will still de displayed. Erase this interference cylinder – you may need a paper space zoom to achieve this?

Inserting the correct support

1 Still with the two (exploded) inserted trays displayed?

2 UCS BASE, lower left viewport active and layer TRAY current.

3 INSERT the red block SUP1 with:
   a) insertion point: 10,0,30
   b) full size with 0 rotation
   c) explode this inserted block.

4 Rectangular array the red support:
   a) for 2 rows and 2 columns
   b) row distance: 80 and column distance: 70.

5 Union the six solids – fig. (e).

Completing the model

1 Create a cylinder with:
   centre: 150,40,0
   radius: 50 and height: 200.

2 Subtract this cylinder from the composite – fig. (f).

3 The model is now complete and can be saved as R14MOD\DESKTRAY.

4 Figure 42.2 displays the four viewport configuration of the model.

5 Before leaving the exercise hide and shade the model. Note the support leg colours – any comments?

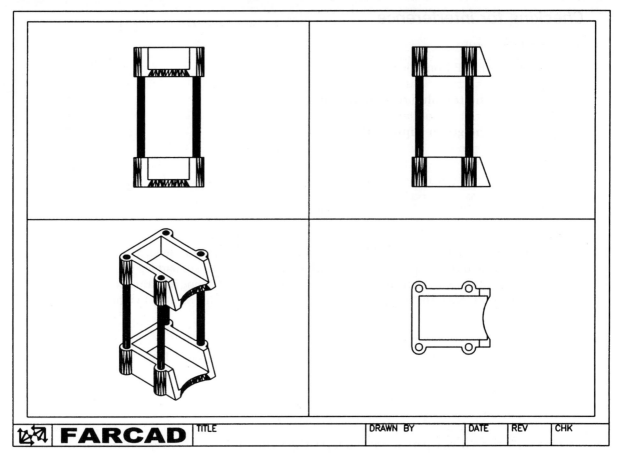

Figure 42.2 Complete desk tray block assembly.

Example 2 – a wall clock

In this example, seven different coloured blocks will be created and used for an assembly drawing. The assembly will then be used for a profile and section extraction.

1 Open your A3SOL template file with layer MODEL and the lower left viewport active. Refer to Fig. 42.3 which displays the blocks to be created.

2 Restore UCS FRONT and zoom centre about 100,75,0 at 250 mag in all viewports.

Creating the blocks

Body (red)

1 Create the clock body from two polylines using the sizes given. The start point should be at 0,0 and your discretion is needed for any sizes not given. Hint: I created the body from lines then used the polyline edit command to join them into one polyline. I then offset the polyline by 6.

2 Solid extrude the two polylines for a height of −20 with 0 taper.

3 Subtract the inside extrusion from the outside extrusion.

4 Create a cylinder with:
   a) centre: −27,125,−10 and diameter: 5
   b) centre of other end option: @54,0,0.

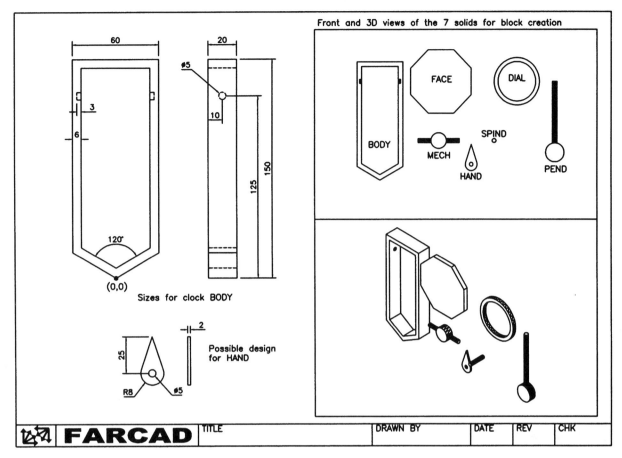

Figure 42.3 Wall clock solids for block creation.

5 Subtract the cylinder from the extruded composite.

6 Make a block of the composite body with:
   a) block name: BODY
   b) insertion base point: 0,0,0.

Face (blue)
1 Draw a blue octagon with:
   a) centre: 80,125,0
   b) circumscribed in a circle of radius: 40.

2 Solid extrude the octagon for a height of 6 with 0 taper.

3 Make a block of the extruded octagon with:
   a) block name: FACE
   b) insertion base point: 80,125,0.

Dial (green)
1 Create two green cylinders:
   a) centre: 180,125,0; diameter: 60; height: 5
   b) centre: 180,125,0; diameter: 50; height: 5.

2 Subtract the smaller cylinder from the larger cylinder.

3 Block the composite as DIAL with insertion base point: 180,125,0.

Mechanism (magenta)

1 Create two magenta cylinders:
   a) centre: 50,50,0; diameter: 5; centre of other end: @54,0,0
   b) centre: 77,50,0; diameter: 20; height: 10.

2 Move the larger cylinder from: 0,0 by: @0,0,−5.

3 Union the two cylinders.

4 Block the composite as MECH, insertion base point: 77,50,0.

Pendulum (cyan)

1 Create two cyan cylinders:
   a) centre: 230,125,0; diameter: 6; centre other end: @0,−90,0
   b) centre: 230,35,0; diameter: 24; height: 10.

2 Move the larger cylinder from: 0,0 by: @0,0,−5.

3 Union the two cylinders.

4 Block the composite as PEND with insertion base point: 230,125,0.

Spindle (colour number 20)

1 Create a cylinder (colour number 20) with:
   a) centre: 150,50,0
   b) diameter: 5 and height: 25.

2 Block as SPIND with insertion base point: 150,50,0.

Hand (colour number 20)

1 Create a cylinder with centre: 120,20,0, diameter: 5 and height: 2.

2 Create a polyline outline using sizes as a guide – own design.

3 Solid extrude the polyline for a height of 2 with 0 taper.

4 Subtract the cylinder from the extrusion.

5 Block as HAND with insertion base point: 120,20,0.

Inserting the blocks

1 Still with UCS FRONT, layer MODEL and lower left viewport.

2 Zoom centre about 0,75,0 at 200 magnification – all viewports.

3 At the command line enter **INSERT <R>** and:
   prompt Block name and enter: **BODY**
   prompt Insertion point and enter: **0,0,0**
   prompt X scale factor and enter: **1**
   prompt Y scale factor and enter: **1**
   prompt Rotation angle and enter: **0**.

4 Insert the other blocks using the following information:

name	insertion pt	scale	rot
FACE	0,125,0	X=Y= 1	0
DIAL	0,125,6	X=Y=1	0
MECH	0,125,−10	X=Y=1	0
PEND	0,125,−10	X=Y=1	0
SPIND	0,125,−5	X=Y=1	0
HAND	0,125,15	X=Y=1	0
HAND	0,125,18	X=Y=0.75	−150.

5   Union the eight inserted blocks – explode needed?

6   Hide, shade, regen then attempt the tasks set for you.

Tasks.

1   In the lower right viewport extract a profile of the model to display hidden detail and remember:
    a) hidden linetype must be loaded and set to the Ph layer
    b) currently freeze layer MODEL in the viewport
    c) optimize the LTSCALE variable.

2   In the top left viewport extract a vertical section through the model and remember:
    a) use the layer SECT with the required section plane
    b) add hatching to the region – UCS is important
    c) currently freeze layer MODEL in this viewport
    d) currently freeze layer SECT in other viewports.

3   Add an additional viewport to display the assembled model from below.

4   The final layout should be as Fig. 42.4 and can be saved.

5   This chapter is now complete.

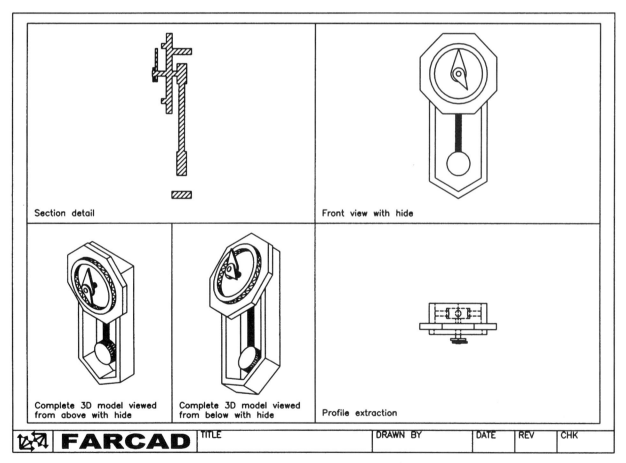

Figure 42.4   Assembled wall clock with profile and section extraction.

Chapter 43

The setup commands

There are two concepts in Release 14 which are new to AutoCAD, these being the Setup View and Setup Draw commands. The two commands can be summarized as follows:

View creates floating viewports using orthographic projection to lay-out multi- and sectional views of 3D solid models.

Draw generates profiles and sections in viewports which have been created with the setup VIEW command.

Basically the two commands allow the user to create multiple viewport configurations which will display top, front, end and auxiliary views as well as extracting profile and sections of the model. In other words the two commands will create the same type of layout that has been achieved with our A3SOL template file, the 3D viewpoint command and with the SECTION and PROFILE commands.

We will demonstrate the two commands with four examples, these being three previously created models and one completely new model.

Example 1 – the backing plate

1 Open the drawing file R14MOD\BACKPLT from Chapter 32 and:
   a) restore UCS BASE with layer MODEL current
   b) in paper space erase three viewports but leave the 3D view
   c) scale the 3D viewport about the point 10,25 by 0.3
   d) in model space, zoom centre about 0,10,50 at 0.3XP.

2 The original model will be centred in this modified viewport.

3 Refer to Fig. 43.1.

The Setup View command

1 Select the SETUP VIEW icon from the Solids toolbar and:

prompt	Ucs/Ortho/Auxiliary/Section/<eXit>
enter	**U <R>** – the Ucs option
prompt	Named/World/?/<Current>
respond	**right-click** – accepting the current (BASE) UCS
prompt	Enter view scale<1>
enter	**1 <R>**
and	paper space entered – icon displayed
prompt	View center
enter	**190,50 <R>**
and	1. UCS icon positioned at the entered point
	2. it is a VIEW UCS
prompt	View center – repositioned view center?
respond	**right-click** – accept entered value
prompt	Clip first corner
enter	**120,25 <R>**

prompt	Clip other corner
enter	**260,85 <R>**
prompt	View name
enter	**TOP <R>**
prompt	Ucs/Ortho/Auxiliary/...
enter	**X <R>** – the end command option
and	a top view of the model is displayed in a viewport.

2 Menu bar with **Draw–Solids–Setup–View** and:

prompt	Ucs/Ortho/Auxiliary/...
enter	**U <R>**
prompt	Named/World/?/<Current>
enter	**N <R>** – the named UCS option
prompt	Name of UCS to restore and enter: **FRONT <R>**
prompt	Enter view scale and enter: **1 <R>**
prompt	View center and enter: **190,185 <R>**
prompt	View center and right-click
prompt	Clip first corner and enter: **120,85 <R>**
prompt	Clip other corner and enter: **260,285 <R>**
prompt	View name and enter: **FRONT <R>**
prompt	Ucs/Ortho/Auxiliary/... and enter: **X <R>**.

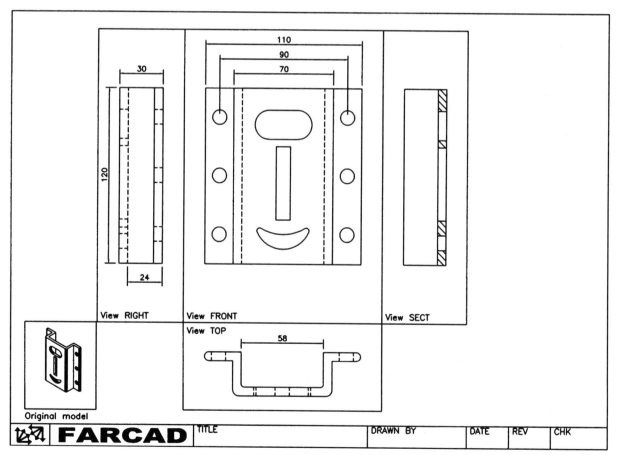

Figure 43.1 View and draw example 1 – backing plate.

3   A front view of the model is displayed in a viewport.

4   At the command line enter **SOLVIEW <R>** and:
     prompt   Ucs/Ortho/Auxiliary/... and enter: **U <R>** then **N <R>**
     prompt   Name of UCS to restore and enter: **RIGHT <R>**
     prompt   Enter view scale and enter: **1 <R>**
     prompt   View center and enter: **90,185 <R>**
     prompt   View center and right-click
     prompt   Clip first corner and enter: **120,85 <R>**
     prompt   Clip other corner and enter: **60,285 <R>**
     prompt   View name and enter: **RIGHT <R>**
     prompt   Ucs/Ortho/Auxiliary/... and enter: **X <R>**.

5   A right view of the model is displayed.

6   Make the new front viewport active and restore UCS FRONT.

7   Activate the setup View command and:
     prompt   Ucs/Ortho/Auxiliary/...
     enter   **S <R>** – the section option
     prompt   Cutting Plane's 1st point
     enter   **0,0,0 <R>**
     prompt   Cutting Plane's 2nd point
     enter   **0,120 <R>**
     and   a dotted 'section' line is displayed
     prompt   Side to view from
     enter   **−10,0 <R>**
     prompt   Enter view scale and enter: **1 <R>**
     prompt   View center and enter: **0,−100<R>**
     prompt   View center and right-click
     prompt   Clip first corner and enter: **260,85 <R>**
     prompt   Clip other corner and enter: **320,285 <R>**
     prompt   View name and enter: **SECT <R>**
     prompt   Ucs/Ortho/Auxiliary/... and enter: **X <R>**.

8   A left view of the model is displayed.

Investigating the layers

1   Menu bar with Format-Layer and note the new layer names:
    a) Front-dim, Front-hid, Front-vis
    b) Right-dim, Right-hid, Right-vis
    c) Sect-dim, Sect-hid, Sect-vis
    d) Top-dim, Top-hid, Top-vis
    e) Vports.

2   Note the layers which are frozen in current viewports and new viewports.

3   Load the linetype Hidden and set the three -hid layers to this hidden linetype.

4   Note: each new viewport created from the Setup View command has had three new viewport specific layers created. These layers are:
    a) -dim: for dimensions
    b) -hid: for hidden detail
    c) -vis: for visible lines.

5   Pick OK when the Hidden linetype has been loaded and set.

The Setup Draw command

1 From the command line, set the following system variables:
   command line   *enter*
   HPNAME   ANSI31
   HPANG   0
   HPSCALE   1.

2 Select the SETUP DRAW icon from the Solids toolbar and:
   prompt   paper space entered
   then   Select viewports to draw
   and   Select objects
   respond   **pick the four new viewports then right-click**.

3 The four new viewports will display the model:
   a) with hidden line removal: top, front, right
   b) as a section: left.

4 Optimize the LTSCALE value.

Task

Add dimensions to the new viewports remembering:
a) the -dim layers are viewport specific but need to be current
b) the correct UCS is needed
c) save the model if required.

Example 2 – the slip block

1 Open model R14MOD\SLIPBL from Chapter 38 and:
   a) in paper space erase three viewports to leave the 3D view
   b) in model space restore UCS BASE and layer MODEL.

2 Use the Setup–View command with:
   a) options: U
   b) name of UCS: current, i.e. BASE
   c) view scale: 0.5
   d) view center: 190,90
   e) clip first corner: 130,50
   f) clip other corner: 250,130
   g) view name: top.

3 In paper space erase the original 3D view then return to model space.

4 Use the SOLVIEW command with:
   a) options: U
   b) named UCS: FRONT
   c) view scale: 0.5
   d) view center: 190,170
   e) clip first corner: 250,130
   f) clip other corner: 130,210
   g) view name: front.

5 Make the new front viewport active and restore UCS FRONT.

6 With the SOLVIEW command use the following:
  a) options: S – section
  b) cutting plane's 1st point: 50,0
  c) cutting plane's 2nd point: 50,100
  d) side to view from: 100,0
  e) view scale: 0.5
  f) view centre: 0,100
  g) clip first corner: 130,210
  h) clip other corner: 10,130
  i) view name: sect.

7 Restore UCS BASE in the top viewport.

8 Activate the Setup View command and:

prompt	Ucs/Ortho/Auxiliary/...
enter	**A <R>** – the auxiliary option
prompt	Inclined Plane's 1st point
respond	**ENDpoint icon and pick pt1**
prompt	Inclined Plane's 2nd point
respond	**ENDpoint icon and pick pt2**
prompt	Side to view from
enter	**–10,–10 <R>**
prompt	View centre
enter	**0,200 <R>**.

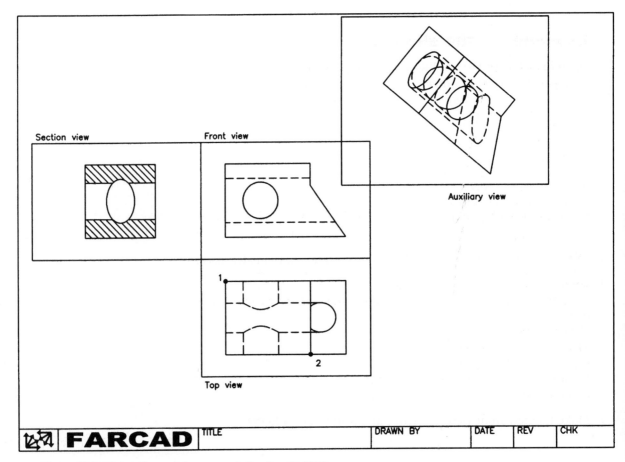

Figure 43.2 Setup view and draw example 2 – the slip block.

prompt   Clip first corner and enter: **375,295 <R>**
prompt   Clip other corner and enter: **230,180 <R>**
prompt   View name and enter: **aux <R>**
prompt   Ucs/Ortho/Auxiliary/...

9   Set the following variables:
    HPNAME: ANSI31
    HPANG: 0
    HPSCALE: 2.

10 Select the Setup Draw icon and pick the four new viewport borders and:
    a) displayed as a section: the left viewport
    b) displayed with hidden line removal: the other three viewports.

11 Save the model as the exercise is now complete.

12 Note: the view centre entry with the section and auxiliary options are perpendicular to the paper space icon. This icon is orientated relative to the:
    a) section plane selected
    b) auxiliary plane selected.

Example 3 – the desk tidy

1   Open the drawing file R14MOD\DESKTIDY created in Chapter 41 and:
    a) paper space to erase the viewports leaving the 3D view
    b) model space with UCS BASE and layer MODEL current
    c) load the linetype HIDDEN
    d) set the LTSCALE variable value
    e) set the following hatch variables:
       HPNAME: ANSI31; HPANG: 0; HPSCALE: 1
    f) refer to Fig. 43.3 for the viewport layouts.

2   Activate the setup View command and:
    a) UCS option with the current (BASE) UCS
    b) view scale: 0.75
    c) view centre: 200,230
    d) clip corners: pick to suit yourself
    e) name: TOP.

3   Erase the 3D view in paper space then return to model space.

4   With the setup View command:
    prompt   Ucs/Ortho/Auxiliary/...
    enter   **O <R>** – the ortho option
    prompt   Pick side of viewport to project
    respond   **pick lower horizontal line of viewport** (snap set?)
    prompt   View center
    enter   **200,150 <R>**
    prompt   Clip corners
    respond   **pick to suit your layout**
    prompt   View name and enter: **FRONT <R>**
    prompt   Options and enter: **X <R>**.

5   Repeat the setup view command and:
   a) select the Ortho option
   b) pick right vertical line of front viewport
   c) view center: pick a point to right to suit
   d) clip corners: pick points to suit
   e) view name: RIGHT.

6   With the setup view command:
   a) select the Auxiliary option
   b) Inclined plane's 1st point: **ENDpoint of pt1**
   c) Inclined plane's 2nd point: **ENDpoint of pt2**
   d) Side to view from: the origin
   e) view center: pick a point to suit
   f) clip corners: position the viewport to suit
   g) view name: AUX.

7   With the top viewport active, select the Section option of the setup View command and:
   a) cutting plane points: pick points as indicated.
   b) side to view from: pick to right of the section line.
   c) view scale: 0.75.
   d) view center: pick to suit.
   e) clip corners: pick viewport to suit the layout.
   f) name: SECT.

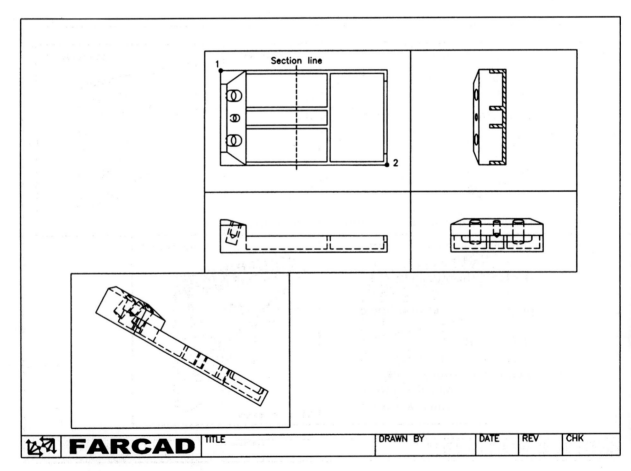

Figure 43.3 View and draw example 3 – the desk tidy.

8 Linetype HIDDEN and three HP variables set?

9 Activate the setup Draw command and pick the five viewports to display the layout with hidden line removal and section detail as Fig. 43.3.

10 *Question*: angle of projection?

11 Save the exercise as it is now complete.

Example 4 – a computer link as a detailed drawing

The three examples selected to demonstrate the View and Draw commands have used models previously created with the A3SOL template file. The fourth example will create a new model 'from scratch', this example being quite complex, but worth the time and effort spent in creating it.

1 In AutoCAD, menu bar with **File–New** and:
prompt   Create New Drawing dialogue box
respond   **pick Start from Scratch–Metric–OK**.

2 Create a new layer MODEL, colour red and current.

3 Set the 3D viewpoint to SE Isometric.

4 Position a UCS origin point to suit and save it as BASE. Remember View–Display–UCS Icon–Origin.

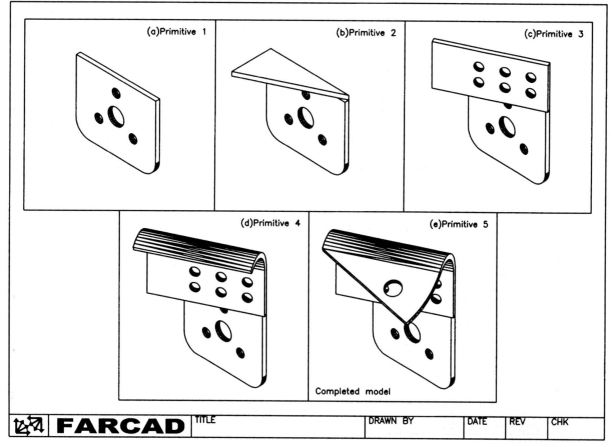

(a)Primitive 1   (b)Primitive 2   (c)Primitive 3

(d)Primitive 4   (e)Primitive 5

Completed model

FARCAD | TITLE | DRAWN BY | DATE | REV | CHK

Figure 43.4   Construction of the computer link model.

5 Set ISOLINES to 18.

6 Set other variables to your own requirements, e.g. units, etc.

7 Refer to Fig. 43.4.

8 The new model will be created from five primitives, each requiring a new UCS position.

Primitive 1: the base

1 Rotate the UCS about the *X* axis by 90 and save as PRIM1.

2 Draw a polyline:
   from: 0,50   to: @60,0   to: @0,–40
   arc to: @–10,–10   line to: @–40,0   arc to: @–10,10
   line to: close.

3 Zoom in on the polyline shape.

4 Solid extrude the polyline for a height of 3 with 0 taper.

5 Create two cylinders:
   centre: 30,25,0   30,40,0
   diameter: 12   6
   height: 3   3.

6 Polar array the smaller cylinder about 30,25 for 3 items with full circle rotation.

7 Subtract the four cylinders from the extruded polyline – fig. (a).

Primitive 2: wedge on top of first primitive

1 UCS PRIM1 current.

2 Set a new 3 point UCS position with:
   a) origin: 0,50,0
   b) x-axis: 60,50,0
   c) y-axis: 0,50,3
   d) save as: PRIM2.

3 Create a wedge with:
   a) corner: 0,0,0
   b) length: 60; width: –3; height: –30.

4 3D rotate this wedge:
   a) about the *X* axis
   b) with 0,0,0 as a point on the axis
   c) for 90 degrees.

5 Union the wedge and the extruded polyline – fig. (b).

Primitive 3: box on top of wedge

1 UCS PRIM2 current.

2 Set a new 3 point UCS position with:
   a) origin: 60,0,0
   b) x-axis: 0,30,0
   c) y-axis: 60,0,–3
   d) save as: PRIM3.

3 Create a box with corner: 0,0,0 and: length: 67.08; width: 30; height: –3. Why 67.08?

4 Create a cylinder with centre: 10,10,0 and: diameter: 6; height: –3.

5 Rectangular array the cylinder:
a) for 2 rows and 3 columns
b) row distance: 10; column distance: 15.

6 *a*) subtract the six cylinders from the box
b) union the box and the composite – fig. (c).

Primitive 4: curved extension on top of box

1 UCS PRIM3 current.

2 Set a new 3 point UCS position with:
a) origin: 0,30,–3
b) *x*-axis: 67.08,30,–3
c) *y*-axis: 0,30,0
d) save as: PRIM4.

3 Zoom in on the 'free edge' of the box.

4 Draw construction lines from: 0,0; to: @0,–15; to: @50,0.

5 Draw a polyline about the 'top rim' of the box using the ENDpoint snap and the close option.

6 With the solid revolve command:
a) objects: enter L <R><R> – to select the polyline
b) options: enter O <R> – object option
c) object: pick the left end of long construction line
d) angle: enter 120.

7 Erase the two construction lines.

8 Union the revolved component and the composite – fig. (d).

Primitive 5: final curved component

1 UCS PRIM4 current.

2 Set a new 3 point UCS position with:
a) origin: 67.08,–22.5,–12.99
b) x-axis: 0,–22.5,–12.99
c) y-axis: 67.08,–24,–15.59
d) save as: PRIM5
e) can you work out the three sets of coordinates?

3 Zoom in of the 'free end' of the curved component.

4 Draw a polyline about the free end of the curved component using the ENDpoint snap and the close option.

5 With the Solid revolve command:
a) objects: enter L <R><R> – to select the polyline
b) options: enter Y <R> – the *Y* axis
c) angle: enter –30.

6 Create a cylinder with:
a) centre: 45,0,15
b) diameter: 10
c) centre of other end: @0,10,0.

7 Subtract the cylinder from the revolved component, then union the revolved component with the cylinder – fig. (e).

8 The model is now complete, so:
   a) restore UCS BASE
   b) save as R14MOD\COMPLINK.

Laying out the viewports

This part of the exercise will involve using all the options of the Setup View command.

1 Model displayed with UCS BASE.

2 *a*) load linetype HIDDEN
   b) set the following variables: HPNAME: ANSI31; HPANG: 0; HPSCALE: 1.

3 Activate the setup View command with:
   a) UCS BASE option
   b) view scale: 1
   c) view center: 180,50
   d) clip corners: 110,10 and 250,110
   e) view name: top.

4 Using the SOLVIEW command:
   a) UCS option with PRIM1 as the named UCS
   b) view scale: 1
   c) view center: 180,200
   d) clip corners: 110,130 and 250,280
   e) view name: front.

5 SOLVIEW command with:
   a) Ortho option
   b) side: pick right vertical side of the front viewport
   c) view center: 60,205
   d) clip corners: 110,130 and 10,280
   e) view name: right.

6 With the setup View command:
   a) activate the Section option
   b) cutting plane points: 30,0 and 30,120
   c) side to view from: origin
   d) view scale: 1
   e) view center: 0,–130
   f) clip corners: 250,130 and 340,280
   g) view name: section.

7 The final SOLVIEW command is with:
   a) the Auxiliary option
   b) inclined plane's 1st point: ENDpoint of pt1 – see Fig. 43.5
   c) inclined plane's 2nd point: PERPendicular to line 23
   d) side to view from: –10,0
   e) view center: 0,210
   f) clip corners: 410,110 and 270,200
   g) view name: auxiliary.

8 Using the Setup Draw command, pick the five viewports to display hidden detail and a section view – linetype HIDDEN loaded?

Tasks

1 With layer model current, draw a rectangle from 0,0 to 420,290.

2 Create a new viewport in the lower left area of the 'paper' to display the model as a 3D view – easy?

3 Interrogate the model:

Area: 19780.44

Mass: 26735.49.

4 Using the viewport specific – dim layers, add the dimensions displayed in Fig. 43.5. Remember you will need to restore the correct UCS for dimensioning. I'll let you figure these out for yourself – especially the auxiliary view dimensions. Note that in my Fig. 43.5 I have accepted the AutoCAD default text and dimension styles.

5 Freeze the Vports layer.

6 In paper space, optimize your drawing.

7 Save the completed exercise – worth the effort?

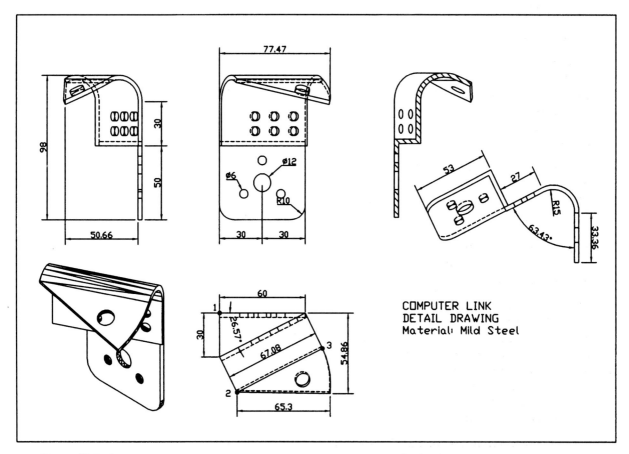

Figure 43.5 The viewport layout created with the View and Draw commands with COMPLINK.

Summary

1   The Setup View and Draw commands are new to Release 14.

2   These two commands allow the user to layout multi-view drawings without the need to create viewports and set viewpoints.

3   Both commands can be activated:
    a) by selecting the icon from the Solids toolbar
    b) from the menu bar with Draw–Solids–Setup
    c) from the command line with SOLVIEW and SOLDRAW.

4   The setup view command has options which allow views to be created:
    a) relative to a named UCS
    b) as an orthographic view relative to a selected viewport
    c) as an auxiliary view relative to an inclined plane
    d) as a section view relative to a cutting plane.

5   When used, the View command creates viewport specific layers, these being relative to the viewport handle number with the following names:
    -vis   for visible lines
    -hid   for hidden lines
    -dim   for dimensions
    -hat   for hatching – only created if the section option is used.

6   The View command requires the user to:
    a) enter the view scale
    b) position the viewport centre point
    c) position the actual viewport (clip) corners.

7   With the Ortho option, both First and Third angle projections can be obtained dependent on which side the new viewport is to be placed.

8   The section option requires that the system variables HPNAME, HPANG and HPSCALE are set. It is usual to use the ANSI31 hatch pattern name, but this is not essential.

9   The Draw command will display models which have been created with the View command:
    a) with visible and hidden detail
    b) as a section if the section option has been used.

10 It is recommended that the linetype HIDDEN be loaded before the View command is used.

11 The hidden linetype appearance is controlled by the LTSCALE system variable.

12 *Note.*
    The user now has two different methods for creating multi-view layouts of solid models:
    a) using the A3SOL template file idea which sets the viewports and viewpoints prior to creating the model. Profiles can then be extracted to display hidden detail
    b) using the VIEW and DRAW commands with a solid composite to layout the drawing in First or Third angle projection with sections and auxiliary views as required.
    c) it is now the user's preference!

Assignment

To give some additional practice with the View and Draw commands I have included another new model. This model is to be created 'from scratch', i.e. in a similar manner to the computer link example. I would suggest the following procedure:

1 begin a new metric drawing from scratch

2 set a 3D viewpoint then complete the model

3 use the View and Draw commands to layout the viewports on an A3 paper

4 add refinements of your choice.

Activity 23: Dispenser container

1 Using the reference sizes below, draw two closed polylines for the dispenser top and body – the UCS position is important. I suggest you make three UCS's – a BASE, FRONT and RIGHT similar to those used in our exercises.

2 Solid revolve the two polylines for a full circle.

3 Create the holes in the top.

4 Use the VIEW and DRAW commands to create a multi-view layout to display:
   a) top, front and right views – with hidden detail
   b) two section views through the planes indicated
   c) a 3D view.

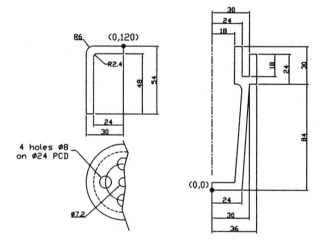

Figure 43.5 Reference details for Activity 23 use discretion for sizes not given.

Chapter 44

Dynamic viewing solid models

The 3D dynamic view command has already been used with other 3D models, but the command is particularly useful with solid models as it allows the model to be 'cut-away' so that the user can 'see inside' the component. We will demonstrate the command with two examples – one new and one previously created.

Example 1 – a nuclear reactor

1 Open your A3SOL template file with UCS BASE, layer MODEL and the lower left viewport active.

2 Set the ISOLINES system variable to 18.

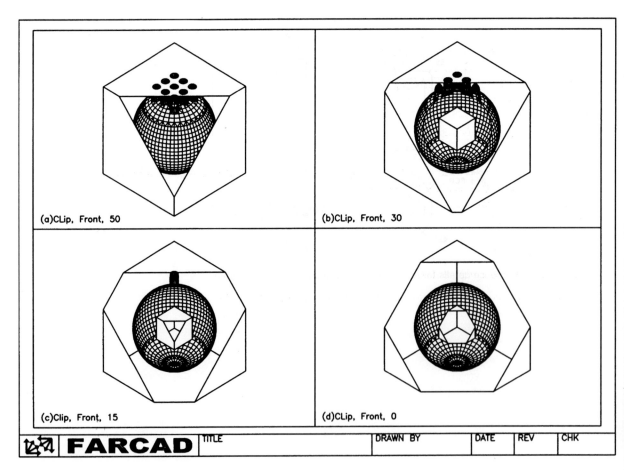

Figure 44.1 Dynamic view example 1 – nuclear reactor.

3 Create the following primitives:

box 1	*box 2*	*sphere*
corner: 0,0,0	centre: 60,60,60	centre: 60,60,60
cube option	cube option	radius: 50
length: 120	length: 30	colour: blue
colour: magenta	colour: red.	

4 Subtract the blue sphere from the magenta box, leaving the red box as a primitive.

5 Create a cylinder with:
   a) centre: 45,45,120
   b) radius: 5 and height: −40
   c) colour: green.

6 Rectangular array the green cylinder:
   a) for three rows and three columns
   b) row and column distances: 15.

7 Subtract the nine green cylinders from the composite – zoom in?

8 Zoom centre about 60,60,60 at 0.6XP in all viewports.

9 Save at this stage as R14MOD\NUCLEAR.

10 Change the 3D viewpoint in all viewports to SE Isometric – need to centre again?

11 With the top left viewport active, select from the menu bar **View–3D Dynamic View** and:

prompt	***Switching to the WCS***
then	CAmera/TArget/...
enter	**CL <R>** – the clip option
prompt	Back/Front/<Off>
enter	**F <R>** – the front clip option
prompt	Eye/ON/OFF/<Distance from target><??>
enter	**50 <R>**
prompt	CAmera/TArget/...
enter	**X <R>** to end command and leave model as fig. (a).

12 Repeat the front clip option of the 3D dynamic view command using the following distance values in the stated viewports:

viewport	*front clip distance*	*fig*
top right	30	b
lower left	15	c
lower right	0	d.

13 Hide and shade the viewports.

14 Save the modified model if required, but not as NUCLEAR.

Example 2 – the computer link

The computer link model was 'created from scratch' to use with the VIEW and DRAW commands in Chapter 43. We will use this model as our second dynamic view exercise.

1 Open the COMPLINK drawing with layer MODEL and ensure UCS BASE is current at 0,0,0 – should be?

2 Menu bar with **Draw–Block–Base** and:
  prompt   Base point<?>
  enter   **0,0,0 <R>**.

3 Now save the model as R14MOD\CL-MOD.

4 Menu bar with **File–New** and select your A3SOL template file with layer MODEL, UCS BASE and the lower left viewport active.

5 Menu bar with **Insert–Block** and:
  prompt   Insert dialogue box
  respond   **pick File**
  prompt   Select Drawing File dialogue box
  respond   1. scroll and pick r14mod folder
             2. pick cl-mod drawing file – preview?
             3. pick Open
  prompt   Insert dialogue box
  with: 1. Block: CL-MOD
             2. File: C;\r14mod\cl-mod
  respond   1. activate explode
             2. pick OK
  prompt   Insertion point and enter: **0,0,0 <R>**
  prompt   Scale factor and enter: **1 <R>**
  prompt   Rotation angle and enter: **0 <R>**.

6 Set the 3D viewpoints in the named viewports as follows:
  top left: NW Isometric   top right: NE Isometric
  lower left: SE Isometric   lower right: SW Isometric.

7 Zoom centre about 30,0,60 at 1XP in all viewports.

8 With the lower left viewport active, enter **DVIEW <R>** at the command line and:
  a) pick the model then right-click
  b) activate the CLip-Front option
  c) enter a distance of 40
  d) enter X to end the command – fig. (a).

9 Repeat the DVIEW command in the other viewports using the **clip front** option with the following distance values:

viewport	dist	fig
top left	25	b
top right	30	c
lower right	0	d.

10 The result should be as Fig. 44.2.

11 The drawing can be saved as the exercise is now complete.

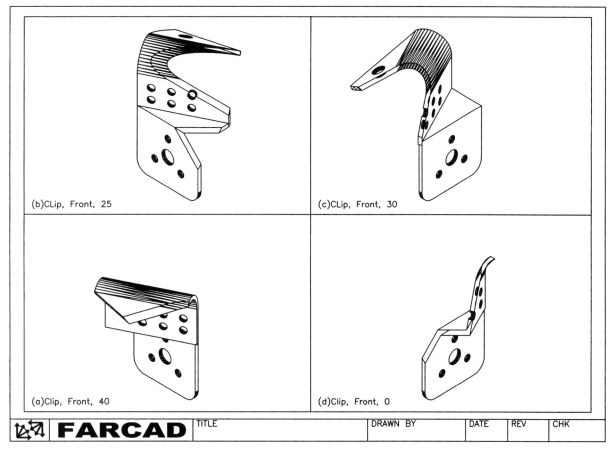

(b)CLip, Front, 25

(c)CLip, Front, 30

(a)Clip, Front, 40

(d)Clip, Front, 0

FARCAD | TITLE | | DRAWN BY | DATE | REV | CHK

Figure 44.2 Dynamic view example 2 – CL-MOD.

Assignment

As this is the last activity in the book I have made it slightly different from the others. You have to use an existing model with the dynamic view command and then add additional views with the VIEW and DRAW commands.

Activity 24: Nuclear reactor

1 Use you drawing file R14MOD\NUCLEAR.

2 Scale and reposition the four existing viewports.

3 Use any four of the dynamic view command in these four viewports.

4 Using the VIEW and DRAW commands set a layout which will display:
   a) a top view with hidden detail
   b) a front view with hidden detail
   c) an end view as a section.

Chapter 45

Finally

If you are reading this I hope that you have worked through the other chapters in the book. If you have, you should now have a good knowledge of the modelling capabilities of Release 14. The progression from extrusions, 3D wire-frames, 3D surfaces and solid models has been deliberate, as I believe this is the correct sequence of understanding modelling techniques.

When Release 14 was first introduced I admit to being a bit wary of what it would contain. The basic modelling tools are all there, but the introduction of the setup View and Draw commands is a great advancement for modelling. No longer does the user have to worry about positioning and centring viewports – the system does this for you. The section and auxiliary options of the View command are excellent, as is the ability to display in first or third angle projection.

Release 14 is a much improved package when compared to Release 13 and (dare I say it) better than Release 12. There are however certain commands which are not in Release 14 but are in Release 12, these being:
a) the ability to extrude/revolve regions to different heights/angles
b) the material properties
c) the SOLCHP command to alter existing model primitives
d) the SOLFEAT command to extract true shapes.

The re-introduction of these features would greatly please the R12 user and make any future AutoCAD release a package hard to beat.

Other concepts AutoDESK should consider are
a) a surface developer
b) a 3D UCS icon
c) realtime 3D movement.

The objective of this book has been to teach the reader modelling with Release 14. I hope that you have enjoyed reading the book and completing the various exercises and activities and as always any comments you have would be more than appreciated. I am always open to suggestions for other ideas for books and for exercises to include.

Bob McFarlane
Bellshill

ACTIVITY 1: EXTRUDED HALF—COUPLING
Using the reference sizes given, construct a
3D extruded model of the half—coupling and
view your model at different viewpoints.

REFERENCE SIZES

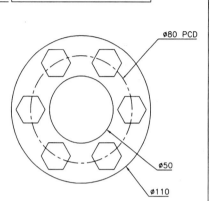

(a) 3D Viewpoint
SE Isometric
No hide

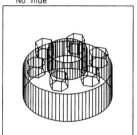

(b) 3D Viewpoint
SE Isometric
With hide

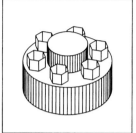

(c) Command line
VPOINT 'R' 300°, −5°
No hide

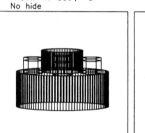

(d) Command line
VPOINT 'R' 300°, −5°
With hide

NB: ONLY 2 'BOLTS' SHOWN IN THIS VIEW

ø80 PCD

ø50

ø110

NOTE: The 'bolts' have been drawn using the
POLYGON command with 6 sides,
circumscribed in a circle of 10 radius
and then polar arrayed for 6 items
with rotation. The centre point of
the array should be obvious?

ACTIVITY 2
Create the 3D wire—frame model
using the sizes given and:
1. set and save a UCS for
   each surface
2. add appropriate text.

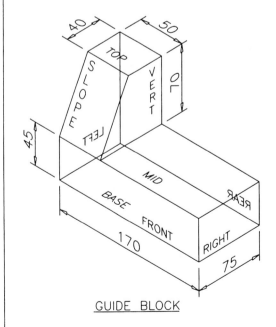

GUIDE BLOCK

ACTIVITY 3
Create the 3D wire—frame model using the
sizes given and add all text with UCS
settings.

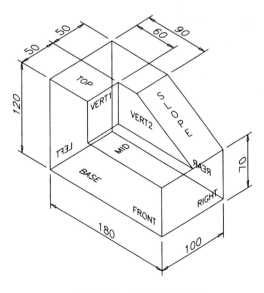

SPECIAL SLIP BLOCK

ACTIVITY 4
(a)Create the shaped block as a wire-frame model
(b)Set and save the five UCS positions
(c)Add the given dimensions
(d)Save as R14MOD\SHBLOCK for future work

MACFARAMUS's SHAPED BLOCK

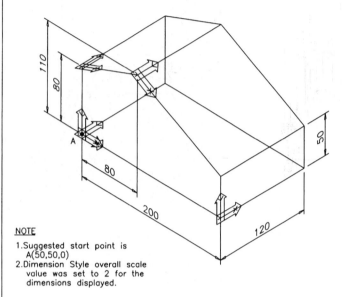

UCS NAMES and ORIENTATION

 BASE

 FRONT

 SLOPE1

 SLOPE2

 RIGHT

NOTE
1.Suggested start point is
  A(50,50,0)
2.Dimension Style overall scale
  value was set to 2 for the
  dimensions displayed.

ACTIVITY 5
(a)Create the wire-frame model using the sizes given
(b)Set and save a UCS for each 'surface' on the
   model. Five UCS names and orientations are
   given.
(c)Add the dimensions
(d)SAve completed madel as R14MOD\PYRAMID

UCS NAMES and ORIENTATION

 BASE

 SLOPE1

 SLOPE2

 SLOPE3

 SLOPE4

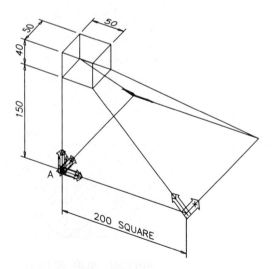

NOTES
1.Suggested start point is A(50,50,0)
2.The viewpoint displayed is with the
  keyboard entry VPOINT R with
  angles of 300 and 40
3.My Dimension Style overall scale
  was set at 2.25

THE SQUARE TOPPED PYRAMID OF MACFARAMUS

ACTIVITY 6
Using the saved UCS settings, add hatching to the four 'surfaces' using the information given.

ACTIVITY 7
Using the slope and vertical saved UCS settings, add hatching to the sloped surfaces, the vertical surfaces and the top surface using the hatch information given.

MACFARAMUS's SHAPED BLOCK

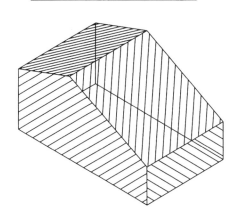

UCS	Pattern	Angle	Spacing
FRONT	user-defined	45	8
RIGHT	user-defined	-45	8
SLOPE1	user-defined	45	8
SLOPE2	user-defined	-45	8

MACFARAMUS's HATCHED PYRAMID

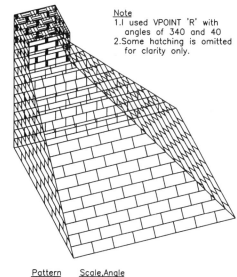

Note
1. I used VPOINT 'R' with angles of 340 and 40
2. Some hatching is omitted for clarity only.

Surfaces	Pattern	Scale,Angle
Four slopes	BRICK	2,0
Four verticals	BRSTONE	1,0
One top	EARTH	-1.5,0

ACTIVITY 8
1. Create a four viewport configuration
2. Display the model at the given viewpoints
3. Centre the model about the point 100,60,55 at 225 magnification.

MACFARAMUS's HATCHED SHAPED BLOCK

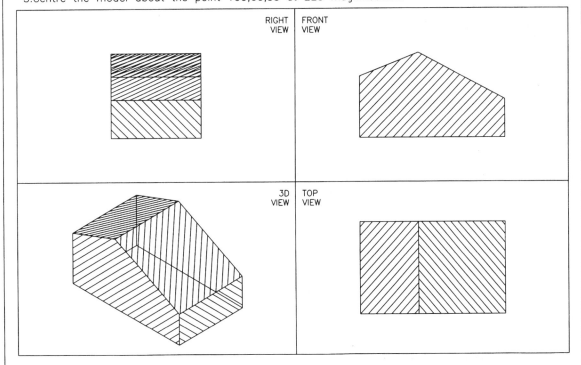

RIGHT VIEW

FRONT VIEW

3D VIEW

TOP VIEW

ACTIVITY 9
1.Create the four viewport configuration
2.Display the model at the given views
3.Centre each viewport about the point 100,100,95 at 275 magnification.

MACFARAMUS's HATCHED PYRAMID

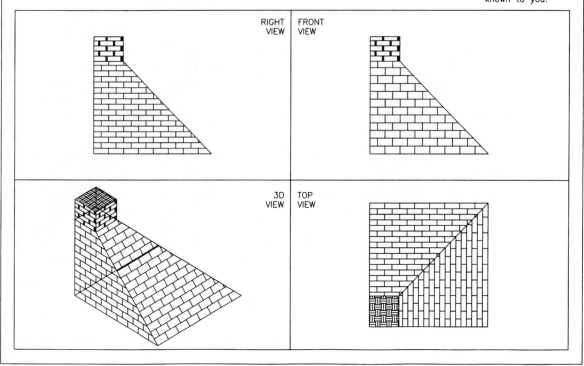

RIGHT VIEW | FRONT VIEW

3D VIEW | TOP VIEW

ACTIVITY 10

Create a wire-frame model of the hexagonal prism using the sizes given and then convert it into a surface model using the 3DFACE command.
Use new coloured layers.

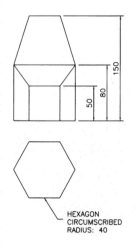

150
80
50

HEXAGON
CIRCUMSCRIBED
RADIUS: 40

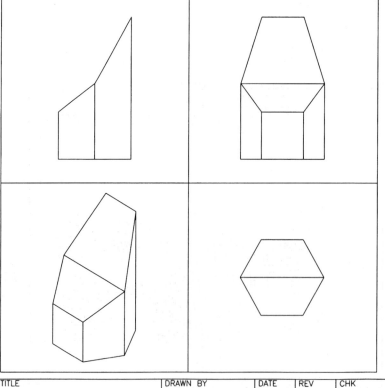

FARCAD

TITLE		DRAWN BY	DATE	REV	CHK

ACTIVITY 11

Using the reference sizes given, create the ruled surface model of the ornamental flower bed. Remember to use layers correctly.

REFERENCE SIZES

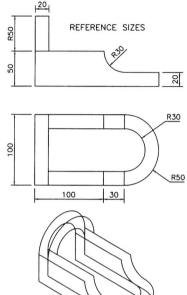

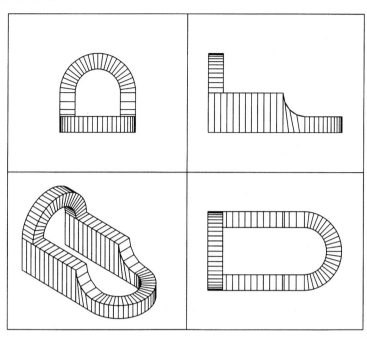

Wire-frame model

	TITLE		DRAWN BY		DATE	REV	CHK
FARCAD							

ACTIVITY 12

Using the reference sizes given, create two path curve profiles and generate the revolved surface for the garden furniture table and chair. The chair has to be polar arrayed for four items.

Note: use your discretion for sizes not given, or design your own furniture.

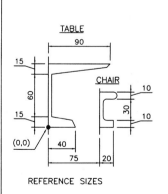

TABLE

CHAIR

REFERENCE SIZES

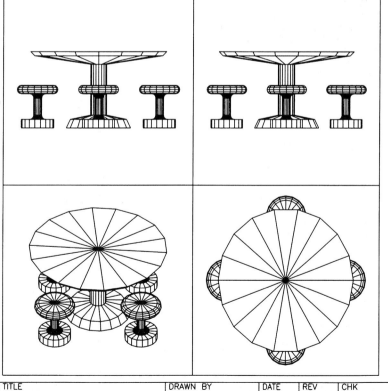

	TITLE		DRAWN BY		DATE	REV	CHK
FARCAD							

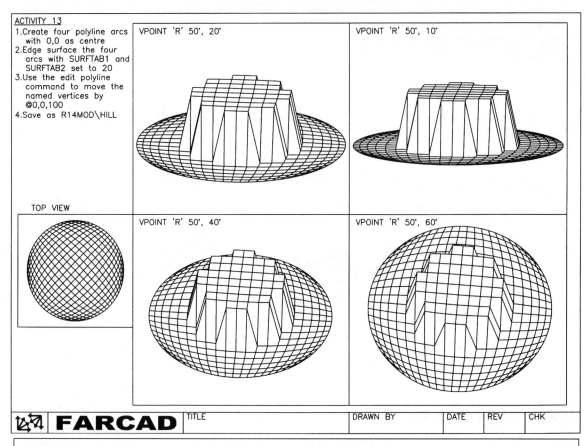

ACTIVITY 13
1. Create four polyline arcs with 0,0 as centre
2. Edge surface the four arcs with SURFTAB1 and SURFTAB2 set to 20
3. Use the edit polyline command to move the named vertices by @0,0,100
4. Save as R14MOD\HILL

VPOINT 'R' 50°, 20°

VPOINT 'R' 50°, 10°

TOP VIEW

VPOINT 'R' 50°, 40°

VPOINT 'R' 50°, 60°

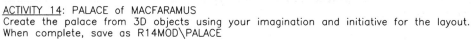

FARCAD | TITLE | | DRAWN BY | DATE | REV | CHK

ACTIVITY 14: PALACE of MACFARAMUS
Create the palace from 3D objects using your imagination and initiative for the layout.
When complete, save as R14MOD\PALACE

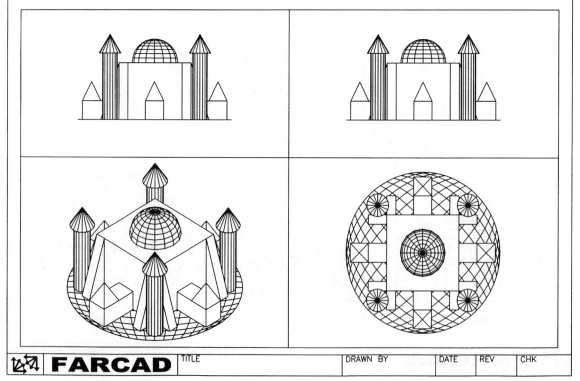

FARCAD | TITLE | | DRAWN BY | DATE | REV | CHK

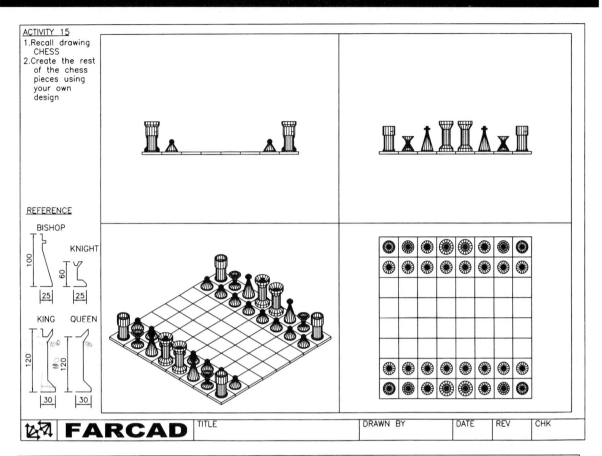

ACTIVITY 15
1. Recall drawing CHESS
2. Create the rest of the chess pieces using your own design

REFERENCE

BISHOP

KNIGHT

100

60

25   25

KING   QUEEN

120

100
120   90

30   30

FARCAD | TITLE | | DRAWN BY | DATE | REV | CHK

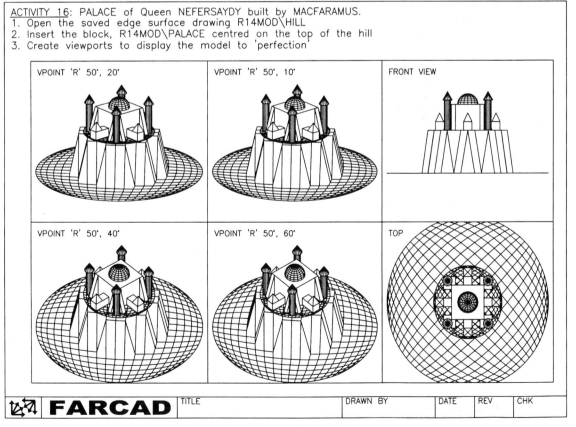

ACTIVITY 16: PALACE of Queen NEFERSAYDY built by MACFARAMUS.
1. Open the saved edge surface drawing R14MOD\HILL
2. Insert the block, R14MOD\PALACE centred on the top of the hill
3. Create viewports to display the model to 'perfection'

VPOINT 'R' 50°, 20°

VPOINT 'R' 50°, 10°

FRONT VIEW

VPOINT 'R' 50°, 40°

VPOINT 'R' 50°, 60°

TOP

FARCAD | TITLE | | DRAWN BY | DATE | REV | CHK

ACTIVITY 17

1. Recall activity 10
2. Make three new viewport specific layers
3. Use the new viewport specific layers to add the dimensions as shown
4. Remember to restore the correct UCS and to make the appropriately named dimension layer current.

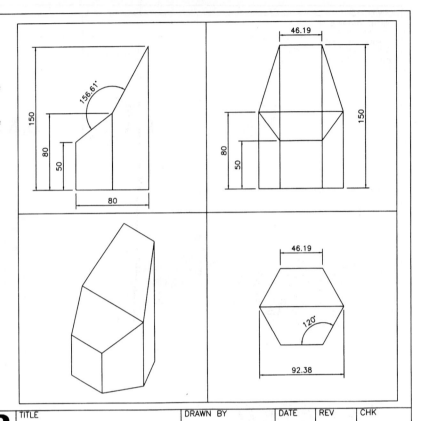

FARCAD	TITLE		DRAWN BY	DATE	REV	CHK

ACTIVITY 18: Create a layout using the 6 solid primitives.

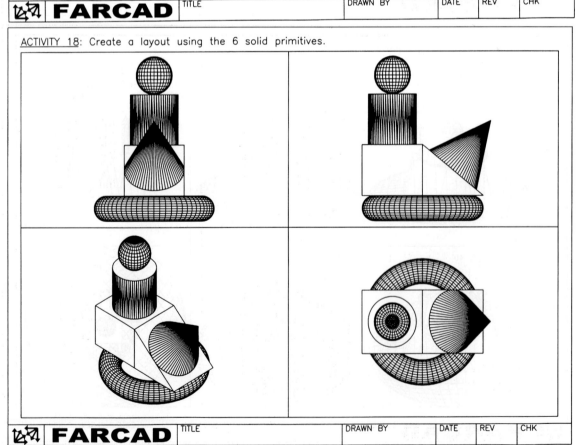

FARCAD	TITLE		DRAWN BY	DATE	REV	CHK

ACTIVITY 19

Create the composite from a cube, box and cylinder. The cube has to be chamfered to give a 'flat top' effect.

Model displayed with hide effect

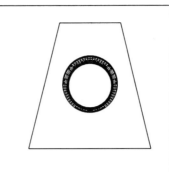
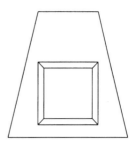
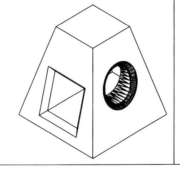
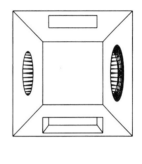

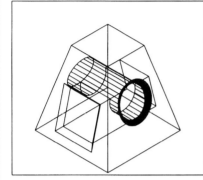

Model displayed without hide

 FARCAD | TITLE | | DRAWN BY | DATE | REV | CHK

ACTIVITY 20

Create a composite model from three extruded regions using the information given. The numbers are 'on the top of the base'.
The extruded heights are:
base: 80
number 1: 50
number 4: 25

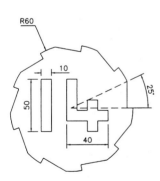

R60
10
50
40
25°

NOTE

Use your discretion for the layout and sizes which are not given.

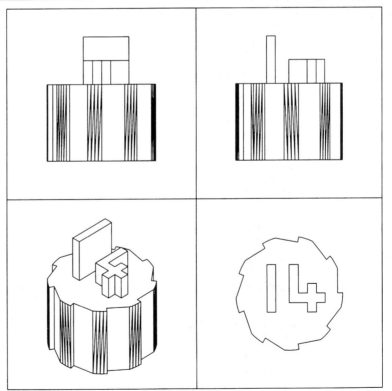

FARCAD | TITLE | | DRAWN BY | DATE | REV | CHK

ACTIVITY 21: CASTING BLOCK
Use viewport specific layers to add the given dimensions.

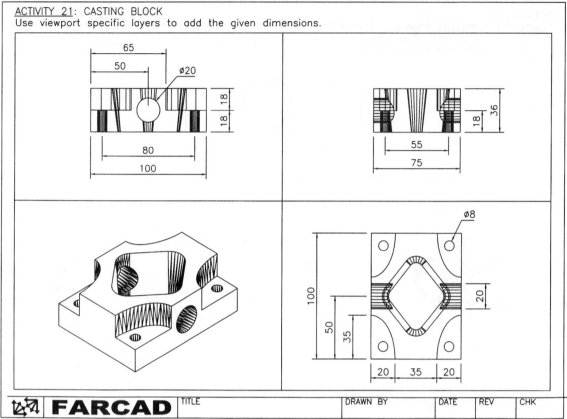

FARCAD | TITLE | | DRAWN BY | DATE | REV | CHK

ACTIVITY 22

Using the model from activity 22 use slice and section on the MODCOMP composite with the planes as shown.

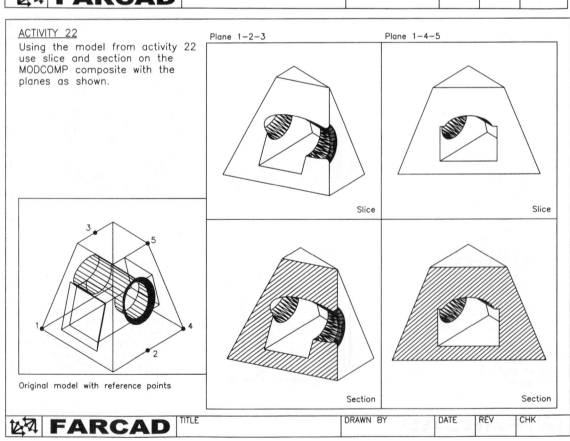

Plane 1-2-3   Plane 1-4-5

Slice   Slice

Section   Section

Original model with reference points

FARCAD | TITLE | | DRAWN BY | DATE | REV | CHK

ACTIVITY 23
Using the reference sizes given, create the model dispenser box, then use
the VIEW and DRAW commands to layout the multi-views as shown

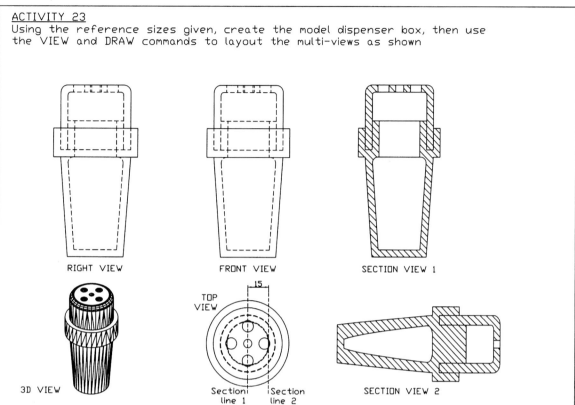

RIGHT VIEW   FRONT VIEW   SECTION VIEW 1

3D VIEW

TOP
VIEW

15

Section Section
line 1 line 2

SECTION VIEW 2

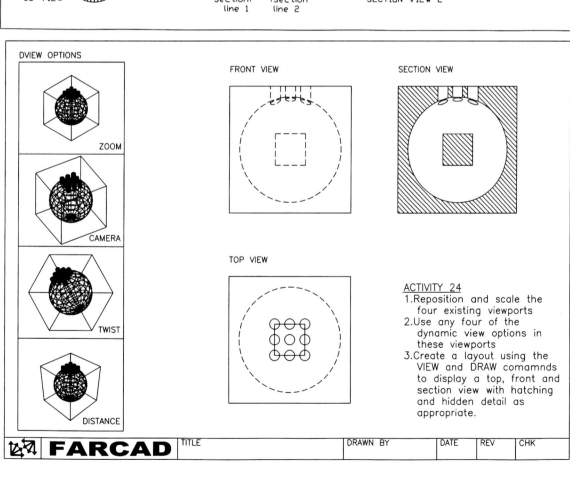

DVIEW OPTIONS

ZOOM

CAMERA

TWIST

DISTANCE

FRONT VIEW

SECTION VIEW

TOP VIEW

ACTIVITY 24
1. Reposition and scale the
   four existing viewports
2. Use any four of the
   dynamic view options in
   these viewports
3. Create a layout using the
   VIEW and DRAW comamnds
   to display a top, front and
   section view with hatching
   and hidden detail as
   appropriate.

FARCAD | TITLE | DRAWN BY | DATE | REV | CHK

Index